DON'T
GIVE UP
THE SHIP

WOKE POLITICS
DON'T
ARE ENDANGERING
GIVE UP
OUR MILITARY
THE SHIP
AND OUR NATION

Calvert Task Group
STRENGTH *through* UNITY

Copyrighted Material

Don't Give Up the Ship
Woke Politics Are Endangering Our Military and Our Nation

Copyright © 2024 by the Calvert Task Group.
All Rights Reserved.

No part of this publication may be reproduced, stored in a retrieval system or transmitted, in any form or by any means—electronic, mechanical, photocopying, recording, or otherwise—without prior written permission from the publisher, except for the inclusion of brief quotations in a review.

For information about this title, contact the publisher:

Calvert Task Group
www.calverttaskgroup.com

ISBNs:
979-8-9908255-0-5 (hardcover)
979-8-9908255-2-9 (softcover)
979-8-9908255-1-2 (eBook)

Printed in the United States of America

Editor: Fred Stuvek Jr.
Cover and Interior design: 1106 Design

Dedication

This compilation of essays is dedicated to the memory of Abraham Lincoln, who served our nation in perhaps the most perilous time in our history. We see parallels between his time in the divisiveness of today's culture over race. These essays were all written by members of the Calvert Task Group in response to what we see as some very real threats to our Constitution. Our stated mission is to support and defend the Constitution of the United States and the United States Naval Academy. We believe that it is our Constitution that makes our nation exceptional. We have all taken an oath to support and defend it, and Lincoln's words inspire us to action.

"We the people are the rightful masters of both Congress and the courts, not to overthrow the Constitution, but to overthrow the men who pervert the Constitution."
—Abraham Lincoln

In Memory of
Rear Admiral Ronald "Rabbit" Christenson, USN (Ret),
U. S. Naval Academy, Class of 1969

Rear Admiral Christenson served our Navy and nation faithfully and with extraordinary dedication that truly made a difference for the better everywhere he served. The Navy well remembers Rabbit's devotion, service, and sacrifice. He will be missed.

Acknowledgment

A special thanks to Fred Stuvek Jr., one of our true patriots, for his leadership in collecting these many essays and assembling them into a defining document that truly represents the motivation for our mission. Fred's total commitment to completing this project underpins our objective to help recapture the true meaning of America as the shining Beacon on the Hill.

Contents

Introduction xv
   The Calvert Task Group—Who We Are xvii
   The Boat School Boys (excerpt) xviii

Starting Point—Where We Come From 1
   Reef Points 3
   The Oath 7
   Honor Points 11
   Truth in Government 15
   A Watershed Event 19
   Shipmates—A Dying Breed 23
   Renaming Military Bases Ignores History 27
   Naval Academy Whitepaper 33
   Where Are the Halseys of Yesteryear? 37

Where We Are Today 41
   The View from 25,000 Feet 43
   Which United States Constitution? 47
   Back to the Constitution 55
   What are DOD's Priorities? 59
   Power 63
   The Tyranny of Time 69

Critical Race Theory in Our Military 73
Critical Race Theory at the U. S. Naval Academy. 77
Diversity, Equity, and Inclusion—or Shipmate . . . You Can't
  Have Both! 79
That Damn Pandemic and the Ultimate School Board 85
Politics and the Academy Boards of Visitors 89
A Monster at the Door 95
The Navy and Diversity 97
Dinner with Joe 105
What Diversity, Equity, and Inclusion Is Doing to the Navy 107
How Social Justice Is Killing the Military 111
How Implementing Diversity, Equity, and Inclusion Will Harm
  Readiness in the Armed Forces and Fail to Solve Anything 115
Divide and Conquer: Radicalizing Military Education 129
The United States: Most Racist Nation Ever 135
DEI Means the Death of Military Professionalism 139
Task Force One Navy Final Report: "The Emperor's New
  Clothes" Redux 145

Why This Matters 161
Now Hear This 163
America's Moral Compass 169
Common Sense 173
America Needs Leaders with Courage 177
You Decide 181
Diversity . . . What Can Be Learned from Fruit Bowl Preparation . . .185
The Navy: Dead in the Water? 189
He Who Will Not Risk Cannot Win 195
A Salute to Our Veterans and a Shared Concern 199

There Is No Justification for the Navy's DEI Push	203
Liberty Matters More	205
A Sailor's Reflections on Race and the Navy	209
The Navy's Misplaced Priorities Versus Core Navy Priorities	213
The Great Diversity Hoax	219
The Navy and Diversity, Equity, and Inclusion	223
Wokeness Is Antithetical to Military Service	229
Danger Close: People's Republic of China	233
Connecting the Dots	237
In Their Own Words: "What Is Critical Race Theory?"	245
Generational Responsibility	255
No Institution Is Safe: Thought Control in the Military	259
Where Do We Go from Here?	**267**
It's Time to Choose	269
The Ultimate Conundrum	273
National Service—A Powerful Woke Antidote	275
The Constitution, the Officer's Oath, Core Values, and Admirals	277
The African Grey Parrot	281
Our Voice Doesn't Matter. Does Our Vote?	285
A Path Forward: For the Navy and the Nation	289
So, Why Are We Here?	295
Conclusion	**299**
A Call to Action for Patriots	299
A Time of Challenge—Moving Toward the Light	301
The Myth of an Apolitical Military: A Call to Action	305
Goodbye DEI, Welcome Back MEI	311
Final Thoughts	313

Authors 315

Appendix 321
Open Letter to Naval Academy Alumni Association 321
Follow-up Letter to Naval Academy Alumni Association 324
Further Reading 329
Glossary 331

Index 333

Notes 347

Introduction

In the summer of 2020, a group of Naval Academy graduates from the Class of 1969 had recently celebrated their 50th graduation anniversary. Close friends, they referred to their golfing brotherhood as the Calvert Group, in honor of Vice Admiral James Calvert, who had been the inspirational Superintendent of the Naval Academy during the years of their attendance. Their conversations had turned from enjoying their retirement years to a genuine concern for what appeared to be a real erosion of values of the institution of the Naval Academy and its mission to develop the leaders of tomorrow. That narrow perspective quickly expanded.

What at first appeared to be a sudden, precipitous, rapid decline in our societal norms, when inspected more closely, reflected an entrenched ideological movement which had been underway for many years. While some apathy might be expected among those enjoying retirement, it became clear that there was a domestic enemy working hard to permanently alter the Constitutional Republic that defines the United States of America.

The conservative majority seemed drowned out by the new vocal rallying points of hate, division, and disruption. The very vocal minority supporting the Movement for Black Lives (AKA Black Lives Matter) issued anti-police and anti-white supremacy edicts, coupled with incessant attacks on the basic tenets of faith, family, and patriotism. Those tactics had their intended effect and are creating divisiveness in our society and military. The new religion of Diversity, Equity, and Inclusion preaches there is no longer a need to judge people on their merit. "Equal Opportunity," long a requirement of law and a hallmark of the American Way, has been abandoned and replaced by "Equity," the quest for equal outcomes distributed on the basis of race and other demographic factors rather than strict merit.

This DEI mania has captured the military despite DOD's own internal surveys consistently showing that racism is considered a problem by only a scant 2% of the force and the fact that DOD's own data shows DOD to be already tremendously diverse.[1]

Could it really be possible that we lost the legacy of the Greatest Generation, the warriors who fought to protect our American value system, in one generation? To quote Ronald Reagan, "Freedom is never more than one generation away from extinction. We didn't pass it to our children in the bloodstream. It must be fought for, protected, and handed on for them to do the same, or one day we will spend our sunset years telling our children and our children's children what it was once like in the United States, when men were free."

Much remains to be done. The sleeping stakeholders of our Constitutional Republic need to wake up. Every congressional, government, military, and judicial leader takes an oath to protect and defend that Constitution. It's time to find out if that oath has any teeth left.

This book is a compendium of articles written by members of the Calvert Group. They are one attempt to increase awareness to the dangerous transition underway with the hope that the American spirit can rally one more time. We hope you find them informative and meaningful.

★ ★ ★

The Calvert Task Group—Who We Are

The Calvert Task Group was formed in 2021 as an association of U.S. Naval Academy alumni, led by members of the class of 1969. Our members believe that American service academies are exceptional institutions for developing future leaders prepared to assume the highest responsibilities of command citizenship and government. Our class motto—*"Non Sibi"* or "Not Self"— appears in the crest inscribed on our class rings and is a common bond that has molded our lives for more than half a century.

We believe we can influence debate and alter the present course of the U.S. Naval Academy, the Navy, and our nation for the better. "Don't give up the ship" is among the most famous of all Navy quotations and reflects the fact that our tradition is courage and persistence even in the face of adversity. We believe that the values of the United States Naval Academy are timeless, and it all begins with the mission. We understand the deep meaning of the Navy's core values of "Honor, Courage, and Commitment" as they apply to everything we do. It empowers our group to defend our ship of state as represented by the Constitution, our Navy, and the United States Naval Academy. We understand that, if we are not part of the solution, we are part of the problem. We are courageous, honorable, and committed and will use those attributes to restore the Naval Academy and the Navy to its foundational roots.

We are now open to all Naval Academy graduates and also welcome others who agree with and are willing to support our cause.

The Calvert Task Group (CTG) represents veterans, retirees, and service-academy alumni, most of whom are combat experienced. We have personally witnessed the unique "color- and gender-blind" culture of our fellow warriors necessary to prevail on the battlefield. Operational team/unit effectiveness requires subordination of self and subgroup identities. It has proven incontrovertible that racial and gender discrimination inherently lowers standards, undermines the chain of command, erodes morale, compromises unit cohesion, and degrades readiness. The result is suboptimal war fighting and diminished mutual trust, causing unnecessary loss of life and, worse, risk of mission failure. Racial diversity as a determining measure of excellence or the key criteria for selection for command is contrary to good order and discipline.

★ ★ ★

The Boat School Boys (excerpt)

By Richard A. Stratton

The United States Naval Academy performs a unique service for the country that other institutions, like my Georgetown and Stanford, never could or should perform. USNA is in the business of forming from the raw material of society a group of leaders of men and women, a class of warriors, a cadre of men and women who are willing to sacrifice their treasure, bodies, and very lives for the Constitution and citizens of the United States of America. USNA recreates the dedication of the signers of the Declaration of Independence, who gave their all for their beliefs. USNA is in the business of developing integrity, honesty, courage, and stamina through rigorous physical and intellectual conditioning.

The product of USNA is not an engineer, a political scientist, a chemist, or a physicist. The product is a citizen, a person formed in an heroic mold, who we hope will never have to be a hero but who we are confident has the fortitude to go into harm's way to protect the Republic. The product is a person who will do the right thing for no other reason than it is the right thing to do. The product is a person who recognizes excellence and is willing to strive for it. The product is a person dedicated to the caring for the enlisted men and women of the U. S. Navy, those people who do most of the work and most of the dying in our Navy. The product is a person that well represents the nation, no matter what port he enters or sea she sails upon. No other institution does this.

The greatest accolade given the USNA in the Vietnamese Communist prison was the statement that the Camp Commander, Major Bui, made to John Sidney McCain III, USNA '58, when John, son of the Commander in Chief Pacific, John, a man born to serve, refused an early propaganda release: "They have taught you too well, McCain! They have taught you too well."

May we always continue to teach the Midshipmen "too well."

★ ★ ★

The Boat School Boys, The New Australian, No. 143, November 29, 1999

Starting Point: Where We Come From

U.S. NAVAL ACADEMY (USNA), PLEBE YEAR. All incoming plebes are given a small, comprehensive book to introduce them to the Navy and the Naval Academy called *Reef Points*. Along with the oath taken on induction day, this little book would guide and inform us as we became formed into Midshipmen learning to become Naval Officers. For most of us, this book's contents remain the cornerstone for our service to our country and our fellow citizens. Plebe year, *Reef Points*, and our common service to our beloved nation made us shipmates, even though we may never have served together after our academy days. There's something special about a shipmate, as we will explain further in these pages.

Learning what we needed to from *Reef Points* was one of our first challenges as plebes. Since then, most of us have lived lives of challenge, both during our time in the service and in civilian life afterward. We were molded in the crucible of the Vietnam War era and the Cold War. Some never made it to civilian life, paying the ultimate price. We have a legacy of warriors beginning with the likes of John Paul Jones and Esek Hopkins at the founding of our Navy, continuing through heroes like Captain Lawrence of "Don't Give Up the Ship" fame, Admiral Farragut ("Damn the torpedoes, . . . Go ahead Jewitt, full speed"), "Hit hard, hit fast, hit often" (Admiral "Bull" Halsey), and so many others. These men were leaders and doers, concerned with winning, not optics and spin politics.

Many things in *Reef Points* were drilled into us during Plebe Year, some of which we were required to memorize. One of the more significant was the "Qualifications of the Naval Officer," once attributed to John Paul Jones but later learned to have been written by Augustus C. Buell in 1900 to reflect his views of John Paul Jones. It speaks of honor explicitly. Honor, trust, and capability are all things that matter. For most of us throughout our careers and our lives, this holds almost as much importance and influence as our oath of office.

It is by no means enough that an officer of the Navy should be a capable mariner. He must be that, of course, but also a great deal more. He should be as well a gentleman of liberal education, refined manners, punctilious courtesy, and the nicest sense of personal honor.

He should be the soul of tact, patience, justice, firmness, and charity. No meritorious act of a subordinate should escape his attention or be left to pass without its reward, even if the reward is only a word of approval. Conversely, he should not be blind to a single fault in any subordinate, though, at the same time, he should be quick and unfailing to distinguish error from malice, thoughtlessness from incompetency, and well-meant shortcoming from heedless or stupid blunder.

The oath of office that we took is a requirement stated in the Constitution and says that we swear to support and defend the Constitution of the United States against all enemies, foreign and domestic. We took that oath seriously and still are bound by it today. And so, in the face of an onslaught that is challenging the Constitution, our nation, and our Armed Forces, we have written these essays to make our voices heard. This chapter will help readers learn who we are and what brings us together. You will understand from our hearts and minds where we are today, why this is important, and why we have some hope and vision for the future.

★ ★ ★

Reef Points

By Jim Tulley

On 30 June 1965, approximately 1300 young men were inducted into the United States Naval Academy class of 1969. We came from all over the country, some with a year or two of college experience, but most right out of high school. Some were the sons of Academy graduates, more from career Navy families, many the sons of Navy veterans, and some came directly from the fleet. Although I don't recall any such discussion at the time, most of us were probably the sons of World War II veterans. Most had been VIPs in a different world, but now we were all Plebes!

We came with a notion of serving our country. The words were on our lips, but at that tender age, it is hard to imagine that we really understood their implication. The Vietnam war had just begun, and it was a time when the faces of graduates who had been killed or were missing in action were on display in the rotunda of Bancroft Hall for all to see. It was a sobering display of what might lie ahead for some of us.

On that life-changing day, each of us received a small book called *Reef Points*. It was sufficiently small to be carried in a pocket, and we were expected to have it with us at all times. It included the Naval Academy Mission, messages from the Superintendent and the Commandant, and pictures of then-modern Navy ships and aircraft. It contained important facts about the campus, called The Yard, and a wealth of nautical tidbits we were expected to memorize and recite to upper classmen at a moment's notice.

Reef Points contained a history of the Academy and quotations from legendary Naval heroes. For most of us, it was our first exposure to the long history and rich traditions of the United States Navy. It was a form of naval indoctrination, but it made us who we are. Fifty-eight years later,

I still have my copy, including little red stars beside all the sections I was expected to know by heart.

Perhaps most important, this little book contained the Code of Conduct, six short paragraphs that described our responsibility in time of war. The Code begins with "I am an American fighting man. I serve in the forces which guard our country and our way of life. I am prepared to give my life in their defense." It ends with, "I will *never* [emphasis added] forget that I am an American fighting man, responsible for my actions, and dedicated to the principles that made my country free. I will trust in my God and the United States of America."

On 4 June 1969, 879 of us graduated and went off to pursue our selected warfare specialties. We were to be designated combat leaders, which in most cases would mean additional training in specific warfighting skills. We were off to a good start and eager to pursue our goals, always mindful that we had signed up to put our lives on the line in defense of freedom and the U.S. Constitution.

Fast forward fifty-four years, and who of us would have thought that we would be in the midst of a new kind of warfare? Who of us could have envisioned the omnipresent dangers inherent in the viruses of Critical Race Theory, Diversity, Equity, and Inclusion, Black Lives Matter, and specious allegations of endemic white supremacy which are infecting our culture? Who of us could have envisioned these cancers metastasizing into our nation, our Navy, and the United States Naval Academy? Who among us could have envisioned such a threat to the survival of our nation as we know it? Instead of the kind of warfare we were trained for so long ago, we are now faced with a warfare of words. It is a warfare for the very soul of the next generation of Americans, and we must recognize that our responsibilities did not end when we hung up our uniforms.

So where do we go from here? Simple—we gather the moral courage to confront this new danger to America and challenge it head-on. We may not be dealing with guns, bombs, and missiles, but we must still use all the smart warfighting skills from our first career as well as new skills obtained since, to fight this national plague. We are still American fighting men, responsible for our actions and still dedicated to the same principles. We are still

American fighting men, and we continue to serve in the forces which guard our country and our way of life. *Non Sibi* (not self) is our class motto, and, in the highest traditions of the Navy, "Don't give up the ship" is our guide.

In the essays that follow, you will read the collective wisdom of more than 535 years of service, honor, and commitment. Fifty years later, the authors are men whose devotion to our country and our Constitution has not diminished but increased over time. Fifty years later, these are men who now more fully understand and appreciate that the oath we all took so many years ago did not end because we took off our uniforms. Fifty years later, these are the beliefs that point our way forward. Finally, a small, inconspicuous book called *Reef Points*, which guided our formative years, now fifty years of life and experience later, points to this, our Belief Points. We humbly ask that you read them with concern and act on them with appropriate courage.

★ ★ ★

The Oath

By Tom Burbage

In June 1965, I raised my right hand in Tecumseh Court along with about 1400 other soon-to-be Midshipmen and took an Oath. While we may not have understood every word of the oath, we took it freely and without reservation. That Oath tied us inextricably to the Constitution of the United States and all it stands for.

Every military officer has taken that same Oath, as have all members of our three branches of Government: Our Congress, the Legislative branch, responsible for enacting the laws that define us, our White House, the Executive branch, charged with executing the laws passed by our Congress, and the Judicial branch, established to settle points of contention. That Oath is to our Constitution, and is taken without exception, every year and with every election.

That Oath is to our Constitution, not to an ideology and not to an Institution. We are a Constitutional Republic. We elect representatives to protect and defend our interests. We are not a pure Democracy based on the collective vote of a population. Pure Democracies tend to have many parties with different interests and tend to form coalition governments composed of several different perspectives to gain a majority. They do not have a constitution to stand behind. They are often short-lived.

In some cases, Democracies eventually lead to dominance of a single individual or party when the Democratic process becomes corrupted because it does not track back to a fundamental charter. Those democracies have historically tended to fail after about 200 years, and we hear many references today that "our Democracy" is doomed to the same fate—it's time for the American ideal to fail.

But the great American experiment to implement a Constitutional Republic is not necessarily subject to the same failure criteria. That may be our fundamental issue today. Is our Constitutional Republic strong enough to resist the suicidal tendencies of other Democracies? Those same tendencies are clearly hard at work within our current administration.

There are clearly forces at work to fundamentally change America. It is difficult to imagine that the impact of open borders will not fundamentally change our American identity. We are a generous nation, despite the many subcultures that seem to want more. Overwhelming our welfare system with the huge influx of illegal immigrants seems to be an objective of the current administration. It is only one of many strategies of the Marxist ideology transformation underway. There are many other examples.

The purity of our election processes has been compromised. It is equally clear that there is a re-emergence of racial discord, proving once again that the financial objective of exploiting racial injustice far exceeds the supply of "white supremacy." Dividing the country along the lines of the oppressed and the oppressors, most conveniently done along racial lines, is well underway.

None of these are aligned with our Constitution. How does that happen? I thought we all took an Oath.

Why are we tolerating the current path our country seems to be on? To many, especially those who have sacrificed much to defend that Constitution, there are many who recited the words of the Oath but viewed it as a non-binding step. There does not seem to be an allegiance to defend our nation against all enemies, both foreign and domestic. Their true oath seems to be to an institution or an ideology, not to our American Constitution.

Most bureaucracies are very slow to change. When they are infiltrated by non-believers, they are even slower. Fundamental American beliefs are now under threat. Faith, family and country, the three fundamentals of our Constitutional Republic, are all under attack. The unity of the American way is now being surgically divided by societal and corporate initiatives incentivized to "Woke up America."

To anyone with an American pulse, the country is on a path to fundamental change, and the nefarious forces at work are evident. The country has been infected, and it has been taking place over many years. Like most diseases, the infection started slowly, but it has now become life-threatening.

So, what do we do, as a nation of patriots? To coin an old phrase, Preach the Gospel, Convert the Agnostics, and Slay the Infidels. Get involved. We are not a traditional democracy, and anyone who succumbs to the gaslighting efforts that it is time for the inevitable reset are not true Americans.

Never forget that we are a Constitutional Republic and that the Oath is clear and unambiguous. Those who take it need to be accountable to defend it. Left unchecked, the American future may well march to a different Constitution—and one that our children and grandchildren will never forgive us for.

★ ★ ★

Honor Points

By Bill Newton

I would like to tell you that I have fond memories of these July Plebe Summer nights, but that would be a lie. What I remember is exhaustion, and sweat running down my back, trying to stay awake in an un-air-conditioned auditorium, watching "Victory at Sea" reruns.

Right now, I want you to pay attention because what I have to say to you is important: important to you personally and important to you professionally. Therefore, if you see your classmates nodding off, elbow them, and wake them up.

1972, Viet Nam: Hanoi was surrounded by Surface to Air missile sites. Our mission was to destroy those sites, to protect the next few day's missions of hundreds of aircrews who would risk their lives hitting targets all over Hanoi.

Flying at 100–500 feet above the deck that night, the anti-aircraft fire was so thick coming up at us that it felt like flying through a 4th of July fireworks display. The seven missiles that they shot at us were so close that they made us veer from our flight plan. In those days, there were no smart bombs, only dumb ones. We missed our target!

What do you say in the debriefing room when your skipper is present, listening, awaiting the news of a success, so he can tell the Air Group Commander, who wants to crow to the Captain, who wants to crow to the Admiral? Will you tell the truth? We did. Our skipper was furious, we caught hell, but the next days' missions knew the truth. Lives were saved.

January 1986, Kennedy Space Center, Florida: A group of engineers were working in the space program. A space-shuttle launch depends on your word.

- NASA and middle management want to launch; bonuses depend on the liftoff going on time.
- Upper management wants to launch. Billions in investment depend on this launch under these specific conditions.
- Most importantly, seven crew members' lives depend on your word.

You know with 100% certainty that the "O" rings on the boosters will fail at the cold temperature prevailing at the time of the launch. Will you tell the truth or fold under pressure?

- They folded.
- The space shuttle *Challenger* exploded, and seven lives were lost.
- I lost my best friend Mike Smith, a '67 graduate, and NASA was set back three years.
- Lack of honor has consequences.

1991, Iraq: General Georges Sada, retired Iraqi Air Force general, was invited by Saddam Hussein to critique a briefing of the plan for the Iraqi Air Force to attack Israel with chemical weapons.

You listen. You process. What do you tell Saddam after the briefing, when you know the mission is a suicide mission, a fool's errand, because of Israel's air superiority? The last general who told Saddam something he didn't want to hear was shot on the spot. Will you tell the truth? Sada told the truth, and God spared his life. The mission never flew.

Will you tell the truth? The answer, if you are a Naval Academy Graduate, ought to be, "You will tell the truth every time!" No matter what!

Why? Because the Naval Academy will spend the next four years teaching you the **habits of a warrior.** And one of the most necessary habits of a true warrior is "Telling the truth, always!"

- If it makes you look good, fine.
- If it makes you look bad, so be it.
- If it hurts, you learn to live with it.
- The truth should and, hopefully, *will* become a part of your DNA.
- A lifelong habit. **A part of who you are!**

The coin that you will get tonight will not make you tell the truth— but it will remind you of who you are—a Naval Academy Graduate—and they tell the truth!

There are thousands of ways to lie, and most of them come to us quite easily. Cheating and stealing are two! Every parent can tell you, *"You don't have to teach a child to lie—it comes naturally!"*

As we age, we become more sophisticated in our lying. Politicians have perfected the art of lying, making it seem as right and natural as breathing.

- Embellishing.
- Failing strategically to remember.
- Omitting critical details.
- Including select points.
- Conveniently forgetting specifics.
- "Spinning" is the current word used for "sophisticated lying."

In the next four years, you will be tempted to engage in these and a thousand other variations of the ancient art of lying, which began with Satan, Adam, and Eve.

The coin given to you tonight has many purposes, but one critical one is this:

- *To cause you to decide tonight, when that coin touches your hand, that when temptation comes, you have willfully decided what you will do. You will tell the truth!*

- If it means getting demerits.
- If it means a lower grade.
- If it means a classmate gets in trouble.
- If it means embarrassment.
- If it means pushups until your arms crater.
- No matter what. You will tell the truth!

You are not politicians. You are not what you used to be. You are Naval Academy Graduates in training, and they tell the truth.

My classmates have built careers on truth. Each has a unique story. Ask them for a story! I know I could trust everyone here with my life, because of their honor and truthfulness. That is a gift greater than silver or gold.

Take the coin. Decide tonight what you will do when you face your temptation. And every time you reach into your pocket and finger that coin, remind yourself why that coin is there. It is there to remind you:

1. You made a decision
2. Who you are
3. You made a decision to always tell the truth; and
4. Who you are. You are a Naval Academy Graduate-in-waiting, and Naval Academy Graduates tell the truth.

May God be with you in your journey!

★ ★ ★

Each year the current plebe class receives a coin the size of a silver dollar, engraved to remind each Mid of the importance of the Honor Concept at the Naval Academy. This essay is the talk given to the incoming class of 2019. July, 2015 Annapolis, Maryland.

Truth in Government

By Brent Ramsey

Today, telling the truth has largely disappeared from public discourse. Let's use basic history as an example. Learning history in school was the norm during the period 1952–1969, the formative years up through my graduation from college. I received a detailed education in public schools of our founding and principles, American history, and world history dating back to antiquity. Part of a Navy family, I went to school all over the country—Virginia, Maryland twice, California, Illinois, and Nebraska. History teaching was consistent everywhere, including its hard truths about slavery here and around the world. After college, I continued studying history in depth, including the Revolutionary period up to our founding, the Civil War period and Lincoln, and the major World Wars in the 20th century, especially WWII.

Studying the founders was particularly insightful and important in terms of really understanding the true nature of America. The founders were highly educated and acutely aware of man's checkered history. They excelled—in extraordinarily challenging circumstances—in establishing a new nation based on principles of liberty and opportunity. Was our founding flawed by slavery and other pre-modern holdovers from a more primitive era? It was. But abolishing slavery at that time was impossible, considering the South's dependence on it as an essential element of its economy. So, evil as it was, had it not been accommodated, America would not have been created at all, and we would likely still be British subjects.[2]

Still, America's founding was and is one of the most extraordinary happenings in the history of the human race. It simultaneously threw off monarchy and created a new form of government with the best elements of all the old forms, a unique new governmental system that had never been tried before. And it worked. America at 248 years is the longest-lived democracy in the world.

A broad knowledge of history and our founding was the expected norm back then for anyone growing up in America. What passes for history today? The execrable 1619 Project, a complete fiction that attempts to paint the picture that our nation was founded on racism.³ Even its author admits it is not history, yet it is being passed off as history to millions of vulnerable young children today, a travesty and a tragedy to deprive the young of today of the truth about our miraculous founding against the longest odds. It is true that slavery survived the founding. But the founders' core principles, as outlined boldly in the Declaration itself that "all men are created equal," created the very foundation for the end of slavery later and the end of racism in 1964 with the passage of the Equal Rights Act. Readers can learn about the lies of the 1619 Project in Peter Woods' *1620: A Critical Response to the 1619 Project* or Mary Grabar's *Debunking the 1619 Project*.

How could such false history have gained such traction in our society? It is because liars are promoting false history for a political agenda to fundamentally change America into a totalitarian socialist state. Lying about our history is the key to overthrowing America. If the majority can be convinced that they are living in an evil country, worse than others and with a system that systematically harms blacks and others, it will be easier to convince them to throw off our Constitution and replace it with socialism. Socialism has been tried, and the result is always death, killing more than 195 million people to date.⁴ Those who refuse to learn from history are doomed to repeat it. We must not go down that path.

Lying about our history is just the start of the lying. Whether it is the mainstream press, the social media giants, academia, Hollywood, corporate America—all these institutions lie about everything. They lie that climate change is a mortal threat to civilization. It's not.⁵ In fact, there has been zero warming for the last eight years. Has the press reported that? They lie that gun violence is out of control. It is not. Americans are safer from guns in 2024 more so than at any time in our recent history, despite a slight increase post-Floyd, when intense pressure from Black Lives Matter caused a dramatic decrease in policing.⁶ Your chances of being killed by a gun (less than 4 per 100,000) is lower by orders of magnitude less than your chances of getting hit by lightning (1 in 15,300).⁷ They lie about abortion being about women's health, when it is really the murder of children in the womb. They lie that the nation has secure borders as millions illegally enter our country, including terrorists and drug runners.⁸ They lie that the rich pay no taxes, despite the U. S. having the most progressive tax regime in the world, where our richest citizens pay a disproportionate percentage of the nation's taxes. For example, in 2018, the top 1% of earners paid 40% of all income taxes, and the top 50% of earners paid 97%.⁹ They even lie about basic biology, exploiting the sad plight of victims of gender dysphoria. In recent years, there has been a 4000% increase in young females claiming to be the opposite sex.¹⁰ Is no one in our vaunted institutions, academia, or the press interested in how this could be happening? They lie about the police being murderous brutes, when the truth is that, overwhelmingly, the police are honest, brave, and fair public servants doing a thankless job to keep our citizenry safe from criminals.

Americans need to demand truth from all our institutions. If a person, a company, an institution is dishonest, they must be challenged by the way you relate to them. How you behave, what you choose to spend money on, the way you consume, what you watch, listen to, or support is everything! If they lie, avoid them. Do not consume their products. Money and interest drive the marketplace. Disney just announced massive layoffs because Americans are staying away in droves from their woke ideology. The power of the purse can restore truth to the public square. Each American must seek truth and use the power of their own pocketbook¹¹ to lift up those institutions, companies, and people who tell the truth and to deny attention and money to those who won't tell the truth. It is up to you!

*This article originally appeared at AmericanThinker.com*¹²

★ ★ ★

Truth In Government, American Thinker. 16 February 2023

A Watershed Event

By Dale Lawson

It was the Summer of '68. Martin Luther King and Robert Kennedy had just been assassinated; there were antiwar protests everywhere; race riots ruled the day. I was on my First Class Cruise from the Naval Academy. I was on the *AGSS 419 Tigeron*, a research submarine. The Cruise is designed as a shakedown for Junior Officers prior to Service Selection and Graduation. I turned 21 and became a Bluenose above the Arctic Circle. The Crew had a funny way of celebrating my drinking-age attainment. They tied a line around me and tossed me overboard. In case you don't know, even in June, the Arctic is *cold*. Fortunately, the ship's Doc gave me a little bottle of the same medicine you find in hotel refrigerators.

After this adventure, we pulled into Londonderry Northern Ireland, which had a Brit sub tender tied up next to the very formal O'Club. A Midshipman from Duke and I had become friends and went out in town. We met a nice young couple, only a couple of years our senior. They were gracious enough to volunteer a tour of the area. All four of us piled into an original Mini, and off we went. We would drive down a street, and the young couple would point out who was Catholic and who was Protestant, which went on for an hour. A pause for a History reminder. The "Troubles" in Northern Ireland didn't start until the following year, so no significant bloodshed was going on. Despite the relative calm, the animus between the two sides was palpable. The two naive back-seaters were incredulous as to the atmosphere. Having never been afraid of placing my foot in my mouth, I finally asked if they could really tell the difference just by looking at their neighbors. The answer was a resounding *"Yes!"* With foot remaining in mouth I then asked . . . "What possible difference does it make?" Then came the ball bat upside the noggin; the wife turned around and said, "What possible difference does a person's skin color make?"

As the title of this essay relays, this was a seminal event in my life. I had grown up in the South, in segregated schools, and the first Black men I went to school with were my two Naval Academy Classmates—and they were the only two. It was just a fact of my life and a sign of the times. The intuitively obvious answer to the Irish woman's question is *None!* We all have our own bias, and I answer to a few, but, for me, this was an important life lesson. The good news is that I learned this while in the uniform of our Nation. All the balderdash of CRT, identity politics, DEI, *etc.,* has no place in our Navy and Marine Corps, as we are one of the most race-blind organizations in this great country. The sole criterion for success is merit—not color. *Hard work* rules the day, not some mysterious manifestation of privilege. I, along with my fellow Brothers in Arms, are heartbroken by what we see happening in our beloved alma mater, our Navy and Marine Corps and the Nation writ large.

Tomorrow is the solemn remembrance of Warriors Lost, Memorial Day. I know they are all in Heaven looking down on us. I pray their vision is degraded so they can't see what has happened to the Nation they Loved and Died for. They, like my Brothers and me, took a sacred Oath to support and defend the Constitution. This Oath does not have a termination date. It is forever and proudly taken. Note that neither the President nor the political parties are mentioned. It is that truly unique idea written nearly two and a half centuries ago by remarkable men . . . Our Founding Fathers. There have been a few additions through the years, but it remains as one of the most important Documents of Human History, along with the Bible and its Ten Commandments, the Magna Carta, and few others. Yet somehow, during this new millennium, we have become a deeply divided nation, with some wanting to destroy the Constitution and start over. Despite the 100% record of failure of Communist states, we have too many folks who want to go that path. We have been here before, in the 1860s. Let us work to ensure a similar result does not take place. Our survival and that of the Free World is in jeopardy.

We face many issues—inflation, an invasion of our Southern border, energy crisis, gun control, abortion, war in Europe, the China situation, on and on. Most of these problems are self-induced by the internecine warfare of the political parties. The party in power seems to operate on the guiding principle of, *Take from one and give to another who is probably less deserving and richer than the first.* The other party is waiting their turn to do the same damn thing. A pox on both their houses! We are being controlled by a marble whorehouse called the Capitol. This has to change. We need statesmen, not politicians. Where are the Washingtons, Jeffersons, Lincolns, TRs, Trumans, and Reagans of our glorious past? We know we have them. Let us each help to bring them forward.

Back to tomorrow. As significant as Memorial Day is, it remains a remembrance of those who have gone before. Through kindred experience with the fallen, I am convinced that each and every hero feels aghast as to how this nation treats its veterans. Whether they are drafted or volunteered, they all answered our country's call. They gave their all with blood, sweat, and tears, and some came home with far less than they started with. Lost limbs, wounded mental health, Agent Orange, burn pits, and the like brought them home to a congress and a nation that only wanted to forget. These brave Soldiers, Sailors, Marines and Airmen are being thrown to the wind. The Veterans Administration is, for the most part, another wasted government boondoggle. Shame on us for allowing this to happen. The men and women of this great nation—who gave up their freedom and health to save ours—deserve better.

★ ★ ★

Shipmates—A Dying Breed

By Brent Ramsey

My Father was a sailor, and my earliest memories are of the Navy, being on bases, seeing sailors in uniform, standing in formation, marching in straight lines, counting cadence, doing the manual-at-arms as one, functioning like a huge living organism with one purpose. As recruit commander, Dad's job was to take raw recruits from all over and to break them down from their undisciplined personal habits, and then build them back up into a cohesive unit of sailors functioning as one. The Navy of the 1950s knew it was necessary to take a bunch of civilians with diverse backgrounds and from all over the nation, with different accents, religious beliefs, educational backgrounds, skin colors, religions, mores, and ethnicities, and mold them into one, into sailors, into shipmates. Navy leadership grasped deeply that they needed to create shipmates, a unified building block of patriotic Americans to carry out defending the nation against our enemies.

What is a shipmate? It is not some random term. It is a term that has a very special meaning. It derives from the immutable truths of the Navy and those "who go down to the sea in ships." Ships and the sea are inherently dangerous things. Those are timeless truths even before you add the extreme dangers created by having enemies. When you go to sea on a ship, you and all the others—literally—are all in the same boat. What happens to one, happens to all. At sea, everyone is aware of the sea state. No one has to tell you what it is. The ship is a living, breathing organism; you feel the ship under you, and you know what it is doing. When at sea, you have to fight the ship for stability at virtually every moment. It is always dangerous . . . unseen hazards, converging ships, adverse weather, navigation challenges, mechanical breakdowns, fleet operations. Every minute, every day, something could happen that will challenge the ship and the crew's safety and even life. To safely and effectively operate a ship takes a unified crew of shipmates—shipmates whose focus is on the ship and what we are doing to carry out our mission.

When you say, "right full rudder," the rudder better come right full, or lives could be lost. Do you think I exaggerate? Constant bearing and decreasing range . . . what to do? Seconds to decide, and lives are at risk. Ask the crews of the *USS McCain* or *USS Fitzgerald*, the ones who are still with us, sailors who will never forget the horror of a collision at sea. And untold thousands who have gone before. The sea is unforgiving, and history echoes with its millions lost there. Add the element of combat or even just rivalry, and the danger rises. This is not the place for social experiment. This is not the place for debate on what should happen next. This is not the place for watching out for hurting someone's feelings or wondering what pronoun to use. I speak from experience as one with thousands of hours at sea on the bridge. I speak from the experience of a dangerous and once highly classified tracking of a Soviet Yankee-class submarine for five days nonstop during the Cold War. I speak from the experience of conning a damaged ship for five days up the East Coast being chased by a hurricane . . . you know that special kind of "fun" of a following sea's action. It takes shipmates, competent shipmates, with unquestioning obedience and quick action to survive those waters. It takes a crew that literally and figuratively are pulling on the same end of the rope.

The social experimentation and woke politics of today's America has no place in the Navy or the other services. It will just get people killed. In the past, when we trained young people who joined, the entire methodology was to forge unity, cohesion, camaraderie, morale, teamwork . . . a group of people who would act with one purpose. Now, we seem intent to categorize and accentuate by race, sex, and other characteristics, and that creates division. At the Academies, where the purpose was to turn out exemplars of professionalism, unity, persistence, toughness, and skill, we now promote affinity groups to divide the whole into competing parts. At Annapolis, thirteen separate affinity groups are listed, including Black Studies, Chinese Culture, and Native American Cultures. A whole month is devoted to "pride" of those attracted to the same sex while ignoring the 70% of the straight, Christian members who abhor what their faith teaches is wrong. That is not how you promote unity. Demeaning the faith of a majority of the military is both wrong and stupid.

The military is governed by the Uniform Code of Military Justice. The operative word is "Uniform." All must be treated the same in matters of merit, behavior, and justice. But that standard is passé. Now, if your skin is the right color, you are glorified. If your sexual orientation is right, you are celebrated. The promotion of what is different about our sailors does not create shipmates. It is wrong thinking and devastatingly damaging to readiness, morale, and retention. The service chiefs have all publicly admitted that recruiting is in crisis. And it will only get worse as normal yearly attrition creates a huge demand each year for new recruits. If you fail to recruit in one year, the problem is twice as big in the next. Now, the services have compounded the problem by separating thousands with religious or other objections to the COVID vax mandate, effectively shooting themselves in the foot for political reasons and ignoring readiness.

To Navy leaders, it's time to create shipmates again . . . before it's too late. *This article originally appeared at AmericanThinker.com.*[13]

★ ★ ★

Shipmates—A Dying Breed, American Thinker, 11 September 2022

Renaming Military Bases Ignores History

By Brent Ramsey

The Department of Defense is studying whether or not to rename military bases that are named for individuals who served in the Confederacy. This article will examine whether or not that is a good idea.

The late great Walter E. Williams, the highly esteemed Black economist, in an article last year at Creators Syndicate, said that "Article 1 of the Treaty of Paris (1783), which ended the war between the Colonies and Great Britain, held 'New Hampshire, Massachusetts Bay, Rhode Island and Providence Plantations, Connecticut, New York, New Jersey, Pennsylvania, Delaware, Maryland, Virginia, North Carolina, South Carolina and Georgia, **to be free sovereign and Independent States.**"[14]

Representatives of these states came together in Philadelphia in 1787 to write a constitution and form a union. During the ratification debates, Virginia's delegates said, "**The powers granted under the constitution being derived from the people of the United States may be resumed by them whensoever the same shall be perverted to their injury or oppression.**" The ratification documents of New York and Rhode Island expressed similar sentiments. At the Constitutional Convention, **a proposal was made to allow the federal government to suppress a seceding state.** James Madison, the "Father of the Constitution," rejected it. The minutes from the debate paraphrased his opinion: "**A union of the states containing such an ingredient (would) provide for its own destruction. The use of force against a state** would look more like a declaration of war than an infliction of punishment and would probably **be considered by the party attacked as a *dissolution of all previous compacts* by which it might be bound."**

If secession had been prohibited by the new Constitution, it would not have been ratified.[15] Thus, naming those who chose to leave the Union and serve the states of their birth and their new creation, the Confederacy, as "traitors" or "treasonous" as many advocates of renaming have done, is not valid, as that possibility of secession by the southern states—and northern states, too—was always anticipated by the framers. Using terms such as "traitors" and "treason" for those who fought for the Confederacy is disingenuous and historically inaccurate. It is remarkable that, today, Robert E. Lee is labeled a traitor guilty of treason such that everything named for him will be replaced. When the Congress, in 1975, restored Lee to citizenship dating back to June 13, 1865, with the House of Representatives voting in favor of his restoration by a vote of 407–10 and the Senate passing by unanimous consent.[16] It ignores our founders' intentions, which clearly envisioned secession as permissible, if not likely, considering the complexity of the situation at the founding.

I would ask anyone in favor of tearing down statues or renaming bases and statues that represent all of our history, as tattered and torn as it is, such as the tragedy of the Civil War, to put yourself in the shoes of those who lived in those troubled times. Imagine you were born in Mississippi in 1825 and you grew up there. You entered the Army or Navy and rose in rank by following the rules and being a good soldier or sailor. You were a military member, not a politician or an historian. Most likely you had little education. You just did your duty for God, state, and country as best you could. Most in this situation were of low rank, but many of the more talented ones ended up as senior officers by the time the war broke out. Then, the unrest began, and things started to unravel. Those who would tear down statues and rename bases and ships would have us believe that the right thing for my notional Mississippian to have done would have been leaving Mississippi at once, traveling north, and joining the Union, leaving his wife and children, his friends, his culture, his very life behind. That strains credulity. Records show that doing such was rare and, from a practical standpoint, nigh on impossible for most. An estimated 1,000,000 men served in the Confederate Army and Navy, most conscripted. Slightly more than 100,000 from the South, the vast majority who lived in close proximity to the Union from either Northern Virginia or Eastern Tennessee served in the Union Army. For any others, there was realistically little practical way they could have served in the Union forces even if they had wanted to, but it was moot, essentially due to able-bodied men's conscription early in the war for a three-year term of service. Are these people to be forever labeled as "traitors" or "treasonous"? Does that really seem right to you?

Let's look more closely at the foot soldiers and sailors of the South. This group was made up of mostly poor, illiterate, disadvantaged whites. Only 3.22% of the whites in the South owned slaves.[17] So, very few actual slave-owners served in the Confederate military ... the vast majority were just poor whites who had little or no choice in the matter. They were victims of circumstances. The poor whites' existence can be described as brutal, desperate, onerous, and short. When the Southern states seceded and decided to fight for their right to leave the Union, they drafted these men by the hundreds of thousands to preserve their economy and their way of life, a way of life they did not invent but one they inherited from their forefathers, and from history. The reality of the history of humanity is that slavery has existed through all recorded human history and still exists today. Were these soldiers and sailors of the Confederacy traitors? Did they have a choice? No, they just did what they had to do. They made an awful choice, one that they had to make, to fight for their state, their culture. It is the only one they knew, and to expect them to have risen above it all and adopted an enlightened view of humanity is not realistic. It's time to stop judging those who rose in rank such that their fame and accomplishments led the leaders who have preceded us to name things after.

In modern times, our national identity is much different than it was in the early years of our country. The national identity was there, but it was subsumed by state and local ties much more so than is the case today. Everything was far different then—technology, travel, communications—everything. We were much more a community of states than a cohesive country. It took two world wars to create the strong national identity that emerged after WWII and that persists to this day. At the time of the Civil War, loyalty to the states was much stronger than to the nation at large, which explains the behavior of many in the Union Army and Navy choosing to resign and join the Confederacy. To reiterate, the concept of secession was not a new one, having existed since the founding, so it was not a surprise that the Southern states chose to do so when faced with economic disaster from the will of the North being imposed on them. This is not an apology for the evil of slavery. It is a realistic acknowledgment of the fiscal reality that Southerners faced: their economic destruction at the hands of the North. As evil as slavery was, the truth was that it was extremely common and accepted all over the world and was part of almost every country's economy. At the time, there were far more slaves in other areas of the New World, such as Spanish and French possessions, or countries who had broken free of their colonial masters much as we had but retained slavery as an economic system. At the time of the Civil War, the slaves transported to America constituted only about 5% of the total slaves brought to the new world, with most concentrated in the Caribbean, Brazil, and other parts of South America.[18] The Civil War was unprecedented. In the history of the world, only one country has ever fought a war to end slavery, the U. S. In 2021, according to *World Population Review*, slavery still exists in more than a hundred countries, and, in six countries, it is still a large fact of life.[19]

General Robert E. Lee has been, up until recently, one of the most admired and even venerated soldiers in our nation's history. He served thirty-two years in the U. S. Army after graduating second in his class and had the distinction of completing four years at West Point, without ever being awarded a single demerit. He served honorably in the Army, fighting bravely in the Mexican War. He was not alone in serving in the Confederate Army, as records show that 151 West Point graduates served as General officers in the Confederate Army. Even though the North and South fought brutal, vicious battles, there is ample evidence that the armies and the soldiers had great respect for each other, and personal animus against those who fought for the South was lacking. When Lincoln and Grant authorized Sherman to sit down and negotiate the end of the war, it was done with civility, dignity, and compassion. Lee and others who fought unsuccessfully for the South were not vilified or condemned by the victors. For more than a century, Lee and others who fought valiantly for what they thought was right in the context of that time and their own cultures and states were honored and admired, held up as exemplars of honesty, bravery, and accomplishment as soldiers and sailors. Only in recent years has history been revised to now view these once-heroes as hateful, racist villains whose memory should be scrubbed completely from the landscape and from our minds. The leaders of the day at the end of the Civil War concluded that it was enough to have won the war and restored the Union to wholeness. Reconciliation and peace were the priority—and the healing of the wounds of war. Now, revisionists and cancel-culture advocates want to re-write the history of the peace at the end and fight the Civil War figuratively all over again. It should not stand. The move to remove Lee and others' names is divisive and unnecessary and serves no meaningful purpose. There is no mandate from Black Americans for this erasing of our history, rather it is an invention of white liberals who are forever trying to stoke division and conflict among us to gain power and control.

I served proudly on the *USS Richard L. Page* (DEG-5). Richard Lucian Page was a lifelong Virginian, entering the Navy in 1824 as a midshipman and serving in many ships and all around the world in our nation's conflicts up until 1861, when the war broke out. He served 37 years of arduous, often dangerous duty, rising to the rank of Commander by the start of the Civil War. When the war broke out, he resigned his commission and joined the Confederate Navy as he saw that as his duty to his home of Virginia. He served bravely in combat in both the Navy and the Army and, by the end of the war, was a Confederate General in charge of the defense of Mobile, Alabama. He and his 400 men fought bravely against a far-superior force of 3,000 Union soldiers and eventually had to surrender when they ran out of powder. He was arrested and spent the remainder of the war in jail.[20] Richard L. Page was a confederate naval and army officer more than a hundred years ago. He served bravely and honorably his entire life, most of that in the U.S. Navy, and I would submit he served honorably in the Confederacy in the context that it was his belief and the belief of millions in the South backed up by the founders' intent that states had a right to dissolve their relationship to the nation when injured or oppressed by the nation. Richard Page was worthy of admiration, and that is why the leaders of the Navy in the 1960s honored him by naming a ship after him. I suspect that the story of each base that is named for a Confederate has a similar story. The bases and the names are part of our history. The fact that a base or ship is named for a famous Confederate is not an endorsement of slavery or racism. Renaming things is part of cancel culture, and I oppose going down that road. It will serve only to divide, and once you accede to the cancel-culture mob, there is no end in sight. This article originally appeared in CD Media.

★ ★ ★

Renaming Military Bases Ignores History, Creative Destruction Media, 12 August 2021

Naval Academy Whitepaper

By Tom Burbage

SERVICE ACADEMIES HAVE HISTORICALLY BEEN unique institutions, with the mission to develop future leaders of our military, government, and industry. The Service Academy culture must be different from other colleges' and universities', due to the expectations of their graduates.

Why is there a growing concern among many alumni today?

Alumni perspectives are heavily influenced by "eras": WWII, Vietnam, introduction of women, Iraq/Afghanistan and now . . . the woke agenda. As time goes on, the perspective of the institution has migrated up that chain and changed. Today's midshipmen were not born on September 11, 2001. What has happened since then, in two decades of relative peace, which has reshaped the focus of the institution?

On paper, the basic mission has not changed: the same oath begins the journey for the new plebes today. This oath, taken at the official swearing-in ceremony, describes the purpose of the U.S. military in a nutshell: to support and defend the Constitution of the United States against all enemies, foreign and domestic.

While many earlier alumni served that oath in combat and many lost their lives defending it, that is likely the first and last time the Constitution of the United States will be mentioned in today's student body.

The demographics of new entrants raising their hands to take that oath has a DNA that has evolved with the society they come from. The institutional decay of our K-12 school systems is embodied in their portrayal of America as an inherently racist nation, white students as genetically bigoted, and minority students as hapless, lifelong victims. The perverse incentives of social programs over that same expanse of time has created a world of fatherless families, with similar effects on the candidate pool.

How does this translate to the admissions process, the first strategic pillar of the Naval Academy mission? It requires new recruiting challenges, with quotas deciding selection, sometimes at the sacrifice of merit. It requires a relaxation of emphasis on the honor concept. The leadership challenge now must deal with an incoming culture with roots in the use and misuse of social media. This is probably the most invasive cultural issue within the Brigade. Unity, teambuilding, and mutual-support concepts so critical to the military ethos are new concepts for many of the incoming students.

Further complicating these challenges are two critical factors. First is a move toward achieving academic ranking with the Ivy League Universities. This has reduced the traditional curricula emphasis on science and engineering to a focus on social engineering, including race, gender, and sexuality immersion. The proliferation of "affinity groups" based on race and ethnic differences again stands in conflict with the "unity ethos" critical to leadership and survival in any combat scenario.

The second is the flow-down of formal policies that support racism runs counter to the Constitution and undermines the *esprit de corps* of military personnel. Parents who are successful in removing these issues from the classrooms in the K-12 system are now seeing these same influences reappearing at the next level. While expected in civilian colleges, it has no place in the Service Academies.

In many ways, the net effect of all of these changes amorphously defines the "domestic threat" the founding fathers feared.

The earlier eras of alumni find it hard to believe, in their twilight years, that the fight seems to have switched from foreign to domestic enemies infiltrating our nation, our society, and our military. Some are unwilling to sit by and watch it happen.

In reality, the domestic-enemy fight has become a real one but . . . is it the only one? Hardly.

We are now facing a new perfect storm. Two near-peer adversaries, war in Europe, an emboldened Iran and their proxies reigniting the war on terror in the Middle East, and elimination of our Southern border are all major and growing challenges. Strategically, in the Indo-Pacific region, an aggressive China has ambitions to rule the world. In many ways, today's graduates have a requirement to prepare for both foreign and domestic threats. It must drive a renewed focus on leadership development.

How are the Naval Academy and our other Service Academies doing? At the Class of 2023 graduation ceremony at the U.S. Air Force Academy, President Biden said, "This year's graduating class is among the most diverse in academy history. You are graduating the highest percentage of women . . . and the highest percentage of minority cadets in history."

A few uncomfortable questions:

1. Is Duty, Honor, Country being replaced by Diversity, Equity, Inclusion?

2. Is the Brigade culture and social-media influence within the dormitory that houses all midshipmen, Bancroft Hall controlled?

3. Does the proliferation of race- gender- and sexual-identity based affinity groups contribute positively to the fundamental military leadership objectives of unity and teambuilding essential to the military ethos? Do Brigade leadership positions within the Brigade reflect merit-based selections?

4. Do Brigade educational lectures outside of the classroom reflect a balanced view for leadership development?

Who has the ultimate concern and responsibility for ensuring that the Mission of the Naval Academy is being achieved? The Board of Visitors? The Alumni Association? Concerned Alumni?

The Board of Visitors is a collateral duty for a mixed group of former Academy graduates and politicians assigned for a specific term of duty. It is the only institution with a specific charter—to evaluate the current state of morale and discipline, the curriculum, instruction, and academic methods, and any/all other matters relating to training midshipmen to lead our Armed Forces in combat—and to make relevant recommendations to the Superintendent. The group meets three to four times a year and receives briefings from Academy leadership. Not intended to be a rubber stamp of progress, it has recently turned political. When the current Administration took over, all three Service Academy Boards of Visitors were terminated, and all previous appointees were summarily relieved of their duties. This act was alleged to be an illegal action and is in court review today.

The Alumni Association is the major fundraising organization that augments federal funding for enhancing improvements to the Naval Academy enterprise. While a noble mission, it has shown no interest in ensuring that the institution of the Naval Academy meets its demanding mission.

To date, those of us with a passion for the mission of the Academy, if we express our concerns, seem to have no place in the "Academy Family." It is time to establish a formal group of concerned alumni who can provide clear and knowledgeable support to the official oversight organization, the Board of Visitors.

★ ★ ★

Where Are the Halseys of Yesteryear?

By Jim Tulley

In the long and storied history of the United States Navy, there have been only four men who attained the five-star rank of Fleet Admiral. All were Naval Academy graduates from a time when that was probably a requirement for attaining the rank of Admiral. What I have been able to uncover from some modest research was that all four shared at least two important qualities. They were all great leaders and all unafraid of taking on the enemies of our nation regardless of military parity.

The biographies of Fleet Admirals William Leahy, Earnest King, Chester Nimitz, and William Halsey are all easily available. In fact, for those who are interested, there is a book, *The Admirals: Nimitz, Halsey, Leahy, and King—The Five-Star Admirals Who Won the War at Sea*.[21] The intent of this paper is not to reiterate their life stories but to review some of what they said, that might be appropriate to today's issues.

To quote Calvert Task Group President Tom Burbage, "We are now facing a much more insidious, hard-to-define, internal threat with the same potential outcome to destroy our freedoms and way of life ... The counterculture has made quiet inroads over our lifetime, expanding without resistance through social media and liberal schooling." What did Leahy, King, Nimitz, and Halsey have to say that might help us in this new war of words?

Admiral Leahy was the oldest and the first to graduate. He graduated in 1897, 35th in a class of 47, served two years at sea as required by law, and was commissioned an Ensign in 1899. He was appointed Chief of Naval Operations in 1937, serving until he retired in August 1939. On that occasion, President Roosevelt said, "Bill, if we have a war, you're going to be right back here helping me run it." In July of 1942, he was called back to active duty as Chief of Staff to the Commander-in-Chief. He was the first, but untitled, Chairman of the Joint Chiefs of Staff.

Admiral Leahy may have been something of a philosopher. On June 5, 1944, in the Commencement speech at Cornell College in Iowa, he said, "Everybody may have peace if they are willing to pay any price for it. Part of this any price is slavery, dishonor of your women, destruction of your homes, denial of your God." Are we willing to pay the price of wokeness? I think not.

Fleet Admiral Earnest King graduated with distinction in the Class of 1901, served two years at sea, and was commissioned an Ensign in 1903. Having had sea duty in destroyers, submarines, and battleships, he transitioned to Naval Aviation. In January of 1927, he reported to the Naval Air Station, Pensacola, for flight training and was designated a Naval Aviator in May of that year. He won his wings as a Captain, at the age of 49. Airplanes were simpler then, but for the Naval Aviators among us, can any of us imagine doing that?

I've discovered two important quotes by Admiral King. He said, "The mark of a great shiphandler is never getting into situations that require great shiphandling." Since we are now in a war of words, our present-day shiphandling is figurative. However, it would seem we have already reached a point in this war that will require great shiphandling. Are we up to the task? He also said, "Nothing remains static in war or military weapons, and it is consequently often dangerous to rely on courses suggested by apparent similarities in the past." It would seem that he is telling us that we must remain vigilant to these new threats and base our actions on what we can do now to eliminate them.

Fleet Admiral Chester Nimitz graduated from the Naval Academy in 1905 and was commissioned an Ensign in 1907. The Naval Academy's yearbook, "Lucky Bag," described him as a man "of cheerful yesterdays and confident tomorrows." Shortly thereafter, he commanded the USS Decatur and was court-martialed for grounding her, an obstacle in his career which he overcame, becoming the third five-star in December 1944. After World War II, he was accorded a hero's welcome at home, but the quiet, self-effacing officer described himself as merely "a representative of the brave men who fought" under his command.

Admiral Nimitz's two most applicable quotations to today's war of words are, "God grant me the courage not to give up what I think is right even though I think it is hopeless" and, "If you're not making waves, you're not under weigh." Some might say that, because of the Washington power structure, the mainstream media, and our "woke" education system, there is little chance to effect meaningful change. I think Admiral Nimitz is telling us not to despair, that we can make a difference, but we'd better be prepared to make some waves.

Fleet Admiral William "Bull" Halsey was undoubtedly the most colorful of the four Fleet Admirals, right down to his nickname. History records that the nickname came, not from his aggressive style, but as the result of a newspaper reporter's typo. It stuck. He attended the Naval Academy Prep School prior to coming to Annapolis, graduated in 1904 and, after his obligatory two years at sea, was commissioned an Ensign in 1906. Prior to taking command of the *USS Saratoga* in 1934, he reported to Naval Air Station, Pensacola, for flight training, and was designated a Naval Aviator in 1935, at the age of 52. Perhaps Halsey's most memorable characteristic was his forcefulness in engaging the enemy and his willingness to take risks to win battles.

Admiral Halsey is unique among the four Fleet Admirals in that he is the only one included in our 1965 version of *Reef Points*, where he is quoted, "Hit hard, hit fast, hit often." Are those words appropriate for our war on wokeness? He also said, "There are no great men, just great circumstances, and how they handle those circumstances will determine the outcome of history," and "All problems become smaller if you don't dodge them but confront them. Touch a thistle timidly, and it pricks you; grasp it boldly, and its spines crumble." In our war of words, I think Admiral Halsey is still speaking to us directly and telling us that any future "greatness" that may be ascribed to The Calvert Task Group and our class will result from our willingness to confront our present circumstances and this new insidious enemy. To quote Abraham Lincoln, "Now we are engaged in a great civil war, testing whether that nation, or any nation so conceived and so dedicated, can long endure." Who could have known? Like Admiral Leahy, we are being called back to duty. Like Admiral King, we will be required to hone our figurative shiphandling, and like Admiral Nimitz, we must have the courage to never give up. Perhaps most of all, like Admiral Halsey, we must act boldly, hit hard, hit fast, and hit often. We are sworn to serve but may continue to ask ourselves, *Where are the Halseys of yesteryear?*

★ ★ ★

Where We Are Today

THE MILITARY TODAY IS BEING TRANSFORMED by the politics of the progressive left. This started in earnest when President Obama issued an Executive Order in 2011 mandating diversity and all its political trappings in the Executive branch of government, including our military. Everywhere we turn, we see the negative effects of diversity, equity, and inclusion (DEI) wherein people are hired or accepted for colleges and universities (and military academies) based on race and ethnicity rather than merit. Entrance exams have been done away with in many cases. Test-score requirements are not applied equally across the board, with more leeway given to favored groups over others, namely Black, Hispanic, and female. Thus, many are selected who are of lesser merit than those who meet the higher standards. Diversity of skin color, ethnicity, gender, sexual orientation, etc., has become more important than diversity of thought and mental or technical prowess. Not only is this morally and legally wrong, but it's also unfair to those with higher qualifications. It is unfair to those being given an unfair leg up because it sets them up for failure in most cases down the line. DEI is extraordinarily damaging to organizational excellence. Instead of emphasizing unity and cohesiveness, traits mandatory for the effectiveness of any organization, it creates disunity, divisiveness, secrecy, envy, and needless internal strife that undermines effectiveness. An ineffective military loses wars and gets people killed. It is also illegal in college admissions, as was decided recently in the *Harvard v. SFFA* and *UNC v. SFFA* cases. It also goes against the Civil Rights Act of 1964, Title VI, and the Equal Protection Clause of the 14th Amendment.

We have previously remained silent, partly because the implementation of many of these policies was hidden. There was a concerted lack of transparency, which continues. In some cases, a sense of not understanding or disbelief or assumption that this was a passing fad led people to be silent. Our generation served honorably and has passed the torch to the generations behind us. Now we see those next generations either silent or duped into thinking DEI is needed or having drunk the Kool-Aid and become true believers. So, we are forced back into battle to save our military and our country.

In this chapter, we discuss some of the things that have prompted us to speak up, to take a stand. We discuss how we got here and how this situation evolved, and what is prompting us to speak out. This is us finding our voice and getting that message out to warn the nation of impending peril.

★ ★ ★

The View from 25,000 Feet

By Tom Burbage

Two clear dynamics are having a corrosive effect on our society—our military and our Service Academies—despite the best intentions to prevent it.

The first dynamic is the intrusion of Marxist-based ideology in the form of Critical Race Theory (CRT) into our public-school systems. This is not an obscure academic concept. It is a politically committed movement that explicitly rejects notions of merit, objectivity, colorblindness, and neutrality of law, among other classically liberal concepts. While there may not be any reason to ban teaching it or any other way of looking at the world, it is dishonest to argue that it is anything less than ideologically radical, intensely racialized, and deliberately polarizing. There is a significant difference between *teaching* alternative theories for academic learning and *preaching* one as a preferred alternative to our Constitutional Republic founding principles. It is even more dishonest to suggest that it exists only in academic cloisters. We live in an era of social engineering being forced into our schools and our government Institutions. These societal factors now influence the candidate pool for our future military leaders. It is a new and unprecedented baggage. Behavioral change must be part of the transition to our military culture, and that transition must occur during the four short years spent at the Service Academies.

The second dynamic is the flow-down of social-engineering experiments into the military. Some view it as the perfect petri dish for experimentation:

a large, captive sample set, subject to government edict and funding, and comprising a diverse slice of American society.

When President Obama issued a directive to integrate women into combat infantry forces, the elite Rangers were forced to lower their physical standards so that women could be admitted. When Marine infantry units integrated women, the male-female units had higher injury rates, slower casualty-evacuation times, poorer marksmanship skills, poorer preparation of fortified fighting positions, and overall lower battle-essential skill sets than all-male units. Although all-male units outperformed coed units in 70% of combat tasks and mixed units were not recommended, this was characterized as "fulfilling the dreams of progressive ideologues at the expense of a service member's life." When the Biden administration took over, the emphasis on wokeism, Critical Race Theory, and reinstating transgender troops increased. We now live in an era of ubiquitous race-based "affinity groups," unfounded allegations of white supremacy or racism when discussing nearly every issue, loose interpretation of law and order and pervasive censorship in mainstream and social media.

As Lt Gen H. R. McMaster, recent National Security Advisor and West Point Distinguished Graduate said, "The U.S. military must continue to evolve toward an institution in which all Americans, regardless of the color of their skin, can fully belong and enjoy equal treatment, because nothing is more destructive to teams than racism or any form of prejudice. But civilian and military leaders must not allow reified postmodernist theories to erode the sacred trust between warriors or diminish the meritocracy and objective realities that are essential to preserving the warrior ethos as the foundation of combat effectiveness. Warriors should be judged by their integrity, trustworthiness, physical toughness, mental resilience, courage, selflessness, and humaneness."[22]

So . . . where should the focus be in this new environment? Better alignment with the new ideology being forced on our society or a renewed focus on developing the historical warrior culture of exceptionally strong leadership. There is an interesting observation by Naval Academy graduate, former Secretary of the Navy, and decorated Vietnam vet Jim Webb that the reality of large-scale war is that the military doesn't fight wars—civilians fight wars . . . but they depend on the best military leadership our country can provide to win. Our veterans who have actually been there and done that hold a strong opinion: that criteria be based on a meritocracy, not a demographic experiment.

How does all this translate to concern among the service-academy alumni? Troubling signs include the following:

1. Proliferation of affinity groups that are based on racial differences.

2. Influence of CRT and liberal teachings in the English and Humanities "college" and the spread throughout the Brigade through lecture series and presentations, including a parade of progressive guest speakers to reinforce the leftward march.

3. Formation of a separate chain of command for Diversity Peer Educators or DEI Representatives at the company level—and the impact of this new structure on the culture of the Brigade.

4. The appearance of an overemphasis on minority selection for leadership positions within the Brigade.

5. The absence of forums for discussions of different perspectives.

6. Sister service academies are experiencing more severe impacts of the new woke environment. Can the Naval Academy be far behind? It is fully dependent on the leadership of the institution.

★ ★ ★

Which United States Constitution?

By Phillip Keuhlen

The choice between the Madisonian Constitution and the Wilsonian "living constitution" is stark and pressing, indeed.

Fifty years ago, my United States Naval Academy Class accepted its individual commissions. Each officer swore an oath of indefinite duration to support and defend the Constitution of the United States against all enemies, foreign and domestic; to bear true faith and allegiance to the same; and to well and faithfully discharge the duties of the office on which they entered. It is an interesting formulation, for it requires one's allegiance to, and defense of, our founding document and, by extension, the principles it embodies.

For virtually all of us, there was clarity about what was meant by "the Constitution of the United States." Even in the unlikely event that a midshipman had arrived at the Academy unschooled in civics, one did not graduate without passing course H303, "U.S. Government and Constitutional Development." Prospective officers understood the Constitution of the United States to be one of the organic laws of the United States, the Madisonian instrument designed by the Constitutional Convention, ratified in 1789, and codified in the Bill of Rights and succeeding amendments, to protect the principles of the American founding stated in the first of the organic laws, the 1776 Declaration of Independence.

One cannot help but wonder *which* United States Constitution today's generation of serving officers intends to support and defend when they swear and live their oaths of office. For, make no mistake, while the Madisonian

Constitution, as amended, and the Declaration of Independence are still the *written* laws of the land, the United States is *administered* today by an entirely different constitutional regime, underpinned by a radically different set of principles. Each officer must face the moral and ethical questions of which to support and defend, and what that defense may entail.

The first principles of the American founding are concisely stated in the Declaration of Independence, the first official act of our Congress and the fountainhead of American law.

> "We hold these **truths** to be **self-evident**, that **all men are created equal**, that they are **endowed by their Creator** with certain **unalienable Rights**, that among these are **Life, Liberty,** and **the pursuit of Happiness**. That **to secure these Rights**, Governments are instituted among Men, **deriving their just powers from the consent of the governed**" (emphasis added).

The political philosophy of natural law and natural rights championed in the Declaration was enshrined by the Congress as our nation's statement of first principles. In the U.S. Code, Congress has placed it at the beginning, under the heading "The Organic Laws of the United States of America," ahead of the Constitution. James Madison, the father of the Constitution, said that it was "the fundamental act of union," the first lawful document by which we illuminate the constitutional principles of Americans.

But, what do the words mean? The founders stated that our rights were individual rights; that they were shared in common by all mankind; that they were conferred by God, not by a state; and that they were ours by right, not rationed by government as a means to its ends. They proclaimed these values incontrovertible, requiring no defense. They declared that the legitimate function of government was to secure the natural rights of individual citizens, and that the power of governments to secure those natural rights derived only from the consent of the people.

The Constitutional Convention labored to craft a document that could be passed out of convention and be ratified by the states. The result was a Constitution founded upon natural law, designed to defend the natural rights of citizens, and equipped with safeguards intended to preserve the union, minimize abuse of power, and assure justice for all. The principal features designed to protect the Republic and constitutional governance included:

- Separation of powers between the legislative (enact law), executive (enforce law), and judicial (interpret law) branches of government;
- Multiple provisions to protect minorities, as individuals, classes, or as states, from the tyranny of a majority by features such as due process, a bicameral legislature, and an Electoral College;
- Formal processes for amending the Constitution;
- Formal processes for electing the president; and
- Reservation of powers not delegated to the United States by the Constitution, nor prohibited by it to the states, to the states or the people, respectively.

The founders' political philosophy of natural law stands in stark contrast to the contemporaneous philosophy of utilitarianism espoused by Jeremy Bentham. Bentham found the motivating principle for society in the shorthand phrase, "The greatest good for the greatest number." This view, rejected by the founders, embraced the state as the arbiter of a "common good"; found right and wrong to be relative, defined in terms of the effect on the "common good"; believed that the "common good," and hence government by the state that defined it, was preeminent over individuals. In utilitarianism, personal liberty exists only to the extent it is bestowed by governments, granted in the context and support of the state-defined "common good."

There is a profound, irreconcilable difference between the first principles of American values stated in the Declaration of Independence and the beliefs of Bentham and his philosophical heirs, such as John Stuart Mill, Georg Wilhelm Hegel, Karl Marx, Herbert Marcuse, and Erich Fromm. Those political philosophers provide the philosophical underpinning of contemporary utilitarian-based Progressive movements. For more than a century, the proponents of Progressivism have unceasingly advanced a vision and values directly opposed to those the country was founded on.

They have mounted sustained legislative, judicial, and executive programs that have incrementally subverted the United States Constitution. They have supplanted it with a regime of governance that compromises essential elements of Madisonian constitutional governance, effectively replacing it with an antithetical political philosophy. And like the proverbial frog cooked in a pan of water brought slowly to boiling, many are totally unaware of this changed constitutional environment.

The seminal political architect of changes that have been impressed into American constitutional governance over the past century was President Woodrow Wilson. Under the cover of the national emergency of World War I, Wilson initiated two radical transformations to Madisonian constitutional governance of limited powers.

- The first, his doctrine of a "living constitution," circumvents provisions of the Constitution by constituting the Supreme Court functionally as a permanently sitting constitutional convention, usurping the power to make law reserved to the elected Congress and inventing new law and constitutional provisions by judicial action, contrary to the Madisonian Constitution's separation of powers.

- Wilson's "administrative state" is an even more profound transformation of constitutional governance. Eliminating the separation of powers entirely, it empowers unelected, unaccountable bureaucratic agencies to reign sovereign over the people, able to make rules with the force of law, enforce them, and adjudicate breaches of them, often absent the due process guaranteed by Madisonian governance.

Wilson's transformations were founded on the Hegelian concept of a state that is sovereign over the people. The Hegelian state functions to define the interests of the community and delimits individual liberty to conform to this revealed state interest. Wilson and his successors have substituted the state as the source and arbiter of citizen's rights, eliminating the Madisonian construct of the state as the servant of the people, governing with their consent in defense of their inalienable rights bestowed from the Creator's fountainhead.

Wilson's Progressive successors have consolidated and extended the transformation of constitutional governance that he initiated. Under cover of another national emergency, the Great Depression, President Franklin Roosevelt radically extended the definition of interstate commerce in Article 1, Section 8 of the constitution. Initially rebuffed by the Supreme Court in his attempt to usurp powers reserved to the states, Roosevelt threatened legislation to pack the Supreme Court with supporters. Thus threatened, the court acceded to Roosevelt's vast expansion of the extraconstitutional regulatory state. This further overturned constitutional protections such as the separation of powers, the presumption of innocence, and standing for judicial review.

Today, the Progressive effort to subvert and functionally replace the Madisonian Constitution and the rights that underpin it continues apace, with current events replete with examples.

- Faced with a thin majority of Supreme Court Justices who profess a nominally Madisonian view of the Constitution and jurisprudence, and who are the sole bulwark against elective tyranny, today's Progressives once again propose to pack the Supreme Court to further consolidate their power.

- Several state legislatures have attacked the Constitution by enacting statutes to enable direct election of the president via the National Popular Vote Interstate Compact. This perversion intends to circumvent several constitutional provisions. These include those for the election of the president; for amendment of the Constitution; and for interstate compacts. At its core, it would disenfranchise many voters by requiring state electors to cast votes based on the voting in other states, rather than in their own.

- Progressives have attacked freedom of speech both directly and indirectly under cover of anti-extremism and fantastic allegations of insurrection, with political indoctrination, cancel culture, speech codes, and de-platforming.

- Prominent Progressives have attacked freedom of religion, perverting the Constitution's intended protection of religious expression. They express open hostility to religion, designed to suppress religious expression and to attack qualification for public office on the basis of professed religious belief. They have initiated numerous government actions to coerce individuals and religious organizations to take actions contrary to their faith.

So, which United States Constitution does the current generation of serving officers support and defend? The formally adopted Madisonian Constitution and its Lockean vision of liberty underpinned in natural law, inalienable rights, and legitimacy based upon the consent of the governed; or the Wilsonian constitutional regime and its Benthamite/Hegelian underpinning in a statist-defined "common good" of contingent rights and liberties?

Will they embrace and fight for the defining values and rights conferred by natural law as enshrined in our founding documents, or abdicate the legacy our forefathers fought and died for and accede to the elective statist tyranny, moral relativism, and legal positivism (judge-made law) of the descendants of Bentham?

Make no mistake; the choice between these visions is *the* critical issue of our American age. Each officer, whether in our individual actions as citizens in civic affairs—or in organizational leadership to our communities, businesses, or government agencies—faces profound moral and ethical decisions in this regard. We will define our place in history and bequeath our greatest gift or curse to our posterity, in the choice we make between them. A significant portion of our body politic unabashedly attacks the vision and values America was founded on, embracing a diametrically opposed vision and set of values, while others simply sleepwalk through it all, content to go along to get along.

From the earliest days of the American experiment, preeminent American leaders have understood how fragile it is. Franklin, asked about the form of the new government after the Constitutional Convention, replied, "A Republic, if you can keep it." Washington warned in his farewell address about the risk of losing shared common vision and values to factionalism. Lincoln, above all, understood and emphasized that preserving America and its common founding vision and values required covenantal rededication by each succeeding generation.

The United States of America has been engaged in a struggle over the choice between our formally adopted Madisonian Constitution and the competing, incrementally advanced Wilsonian constitutional regime for more than a century. Taking inspiration from Patrick Henry's speech before the Virginia Convention at St. John's Church in Richmond on March 23, 1775, we argue that those who believe in America's founding values but deny this reality have eyes but see not; have ears but hear not. Hearkening back to a time of youth perhaps, they cry for a return to civility and restraint in public discourse, crying "Peace, peace!"—but there can be no peace between these alternatives.

We are potentially at a tipping point in the history of our Constitution and constitutional governance. Those who crusade for Wilsonian governance and its rejection of America's founding values are today ascendant in government institutions and much of public culture. They already display the inclination to suppress opposition by such means as are available to them, including political indoctrination and broad proscription/punishment of "political speech" in the Armed Forces. Those who take an oath to support and defend the Constitution of the United States of America face choices in the execution of that responsibility.

What does it mean to support the Constitution? Is their obligation passive or active? Does military service alone fulfill the obligation to support and defend the Constitution? Is there an obligation to speak up publicly, either inside or outside the military organization to support the Constitution? Is such speech a "political" or "extremist" activity, or is it a core element of carrying out their oath of office? All are fair questions, yet all must be preceded by officers first consciously answering for themselves, "Which Constitution of the United States am I defending?" The choice between the Madisonian Constitution that remains the law of the land or the Wilsonian "living constitution" that subverts its intent, denies its underpinning values, and is increasingly the basis of current governance is stark and pressing indeed.

★ ★ ★

"**Which United States Constitution**," American Greatness, 17 October 2021

Back to the Constitution

By Tom Burbage

What if we called a timeout and went back to the Constitution?

There are many positions being tested against the fabric of our nation. Many are clearly taken from the Socialist teachings of Karl Marx and his modern-day mouthpieces. This is a distributed bunch, focused on the many pieces of the puzzle that have the potential to change a nation within a single generation. They are not unique to this Administration, but they have found a soapbox within this Administration that has moved their voices from the fringe to the mainstream.

Today, real history is seldom taught in our school systems. After years of minimal interest in school curricula, should we be surprised that our public schools have become one of the key breeding grounds for the new vision for America? Awareness of this travesty was suddenly elevated with the virtual learning resulting from the pandemic. Parents woke up. Racial-justice warriors, who operate for personal gain and have done little for the real issues that affect our society today, have been plying their trade for decades.

How many students today know that the United States moved from Articles of Confederation, adopted in 1777, to a Constitution, ratified in June 1789? The Constitution established the three branches of government, bi-cameral chambers of Congress and an executive branch headed by the President to implement the decisions of Congress.

A written constitution underscores the fact that words matter. In our Constitutional Republic, elected officials at all levels of government and many others take an oath to support and defend the Constitution of the

United States as it is written, not arbitrary interpretations of its intent. Many federal, state, and local employees, as well as all members of the military, take the same oath.

So, why is there such a divide in the country today? As Scott Bradley said in the quote above, words matter. It is easy to define a foreign enemy. They tend to be defined by lines and uniforms. The Iron Curtain, The Berlin Wall, the 38th Parallel, the Maginot Line, to name a few. They have distinct characteristics and intentions that most can agree on. Our entire military defense structure is built on keeping America safe from foreign enemies.

Domestic enemies have none of those. They attack through fundamental institutional change. We established the Department of Homeland Security as a new element after 9-11. We have all witnessed the ease of quickly neutering that organization with the debacle on the Southern border. Domestic enemies also have supporters within our institutions who would not agree with that moniker, which contributes to the gridlock in developing any significant response.

The perfect storm of national decline may be the convergence of four things:

1. the imposition of critical theory, both law and race. Lady Justice is no longer blind, laws are unevenly applied or ignored completely. Indoctrination in race-based ideology separates oppressors and the oppressed based solely on the color of their skin. Both have slid under the radar and now have traction.

2. the change in our public-school curricula. What is now being taught is just as bad as what is not now being taught. Revisionist history attacks the Constitution as our guiding principle. But what is our Constitution? These are the schools delivering the next-generation candidate pool for the leadership positions of America.

3. The forced movement to race-based indoctrination in our military, the one institution that has been and should always be merit-based. Originally labeled Critical Race Training (CRT), it then changed to Diversity, Equity, and Inclusion (DEI). Words matter, and changing words does not change intent.

4. ==Lack of will to honor the oath all of us== took as military veterans, members of Congress, and political leaders to at least question the impact of this new *domestic threat.*

Is it time to think about the Oath that binds us?

★ ★ ★

What Are DOD's Priorities?

By Brent Ramsey

Faced with the burgeoning threats from China, Russia, Iran, North Korea, and radical Islam, what priorities does DOD have? Below is a summary of recent DOD and service activity that appears unrelated in any obvious way to readiness and the defense of the nation:

- DOD Stand Down ordered on 5 Feb 2021 directing an all hands Stand Down, focusing on white supremacism in the military. 1.4 million service men and women were ordered to participate. No evidence of widespread white supremacists or extremism in the military has been provided by DOD. In fact, after the stand-down, DOD reported fewer than 100 extremism incidents in a force of 2.1 million.[23]

- In June 2020, the Navy stood up Task Force 1 Navy to promote diversity, inclusion, equity, and belonging.[24] The Navy's effort is in support of the DOD Board of Diversity and Inclusion that was formed to address those same issues. There is strong evidence that Critical Race Theory (CRT) is being taught at the Naval Academy. How do diversity, equity, inclusion, belonging, or CRT training improve combat readiness?

- On April 9, 2021 the Secretary of Defense announced the creation of a "combating extremism" working group and has given the group 90 days to come up with recommendations. "Why the emphasis?" asks Congressman Mike Rogers, who cited statistics showing that only nine Army soldiers have been separated for extremism activity in recent years out of a force of 1.1 million.

- On March 31, 2021 Secretary Austin signed the following statement: "On this International Transgender Day of Visibility, we recognize the great strides our Nation has made raising awareness of the challenges faced by the transgender community. Their shared stories of struggle remind us that more work needs to be done to ensure that every person is treated with dignity and respect no matter how they identify. To that end, I am pleased to announce we have updated departmental policy governing the open service of transgender individuals in the military. This updated policy reinforces our previous decision to allow for the recruitment, retention, and care of all qualified transgender individuals. It also allows for a short implementation period for the military departments and services to update their policies. The United States military is the greatest fighting force on the planet because we are composed of an all-volunteer team willing to step up and defend the rights and freedoms of all Americans. And we will remain the best and most capable team because we avail ourselves of the best possible talent that America has to offer, regardless of gender identity."

- How is having transgendered individuals in the military improving military readiness and our ability to fight and defeat our enemies?

- CNO Reading List includes the following books, recommended for all sailors to read. After much controversy, the CNO removed the books from the recommended list:[25]
  - *How to Be an Anti-Racist* by Ibram X. Kendi. This book promotes CRT and has been widely criticized as being racist.
  - *The New Jim Crow* by Michelle Alexander. The book's premise is that the reason that there are millions of Blacks in America's prisons is because the justice system in the U.S. is a deck stacked in a racist fashion against Blacks, and the high incarceration rates for Blacks is because we are a racist nation, not because Blacks commit a huge amount of the crimes, disproportionate to the percent of the

population that they represent. DOJ's Bureau of Justice Statistics reports document that, despite being only 12.5% of the population, Blacks commit 35% of the violent crimes, which explains their high prison population. How does a book with doubtful validity based on DOJ statistics land on the CNO's recommended reading list for all sailors?

- *Sexual Minorities and Politics* by Jason Pierceson. *Sexual Minorities and Politics* is a history of how the LGBT movement changed the law in America regarding marriage. How do sailors reading about how marriage law changed help the Navy become a more effective fighting force?

- The terms "diversity," "equity," "critical race theory (CRT)," "intersectionalism," and "lived experience" are found in frequent use in today's military. What is the origin of the use of these words, and what does their use and the implementation of their concepts mean for the readiness of our military forces? Organizations such as the Heritage Foundation, prominent scholars at the Manhattan Institute (Heather MacDonald), the Hoover Institution (Drs. Thomas Sowell and Shelby Steele), the esteemed Civil Rights activist, scholar, and author Mr. Robert Woodson, and scholar, author, and professor Dr. Wilfred Rielly warn that these words and concepts are dangerous and contrary to our founding constitutional principles and are actually the opposite of what they claim to be.

- There is a tremendous emphasis on race in today's military. Is this an appropriate use of time and resources in a constrained environment? An extensive analysis done by Christopher Rufo for the Heritage Foundation found evidence of **class** supplanting **race** as the determinative factor for success of Americans of all races. Quoting from the analysis:

"According to a growing body of evidence, social class is gradually supplanting race as the most salient variable for producing inequality. With regard to family, as Harvard scholar Robert Putnam has observed, '[t]he class gap over the last 20 years in unmarried births, controlling for race, has doubled, and the racial gap, controlling for class, has been cut in half.[26] With regard to workforce participation, Census data show that black and white Americans with the same educational attainment have roughly equivalent levels of workforce participation: Education as a proxy for class has a much greater impact than race.[27] With regard to education, Stanford professor Sean Reardon shows that the class gap in academic achievement is 'now nearly twice as large as the black-white achievement gap,' in contrast to a half century ago, when 'the black-white gap was one and a half to two times as large as the [class] gap.'"[28] The data is clear: blacks in America can avoid the poverty trap if they get married before having children, they graduate high school, and they are employed full-time.[29]

Questions:

1. Will the DOD emphasis on the above issues improve military readiness?

2. How will the emphasis on the topics above prepare our soldiers, sailors, airmen, or marines to defend the nation against our enemies? The GAO released a report on 7 April 2021 that documents falling readiness across the board in the military.

3. How does the DOD emphasis on racism, extremism, transgenderism, diversity, and CRT improve readiness?

Conclusion: Organizations and citizens should engage the national leadership on these issues to urge DOD to examine the validity, urgency, and necessity for such programs. *This article appeared in CD Media and Real Clear Defense.*

★ ★ ★

What Are DOD's Priorities? Real Clear Defense, 6 June 2023

Power

By Guy Higgins

I'VE BEEN THINKING ABOUT THIS. Lord Acton observed, "Power corrupts, and absolute power corrupts absolutely." In the summer of 1967, while playing Marine at Little Creek, I read Frank Herbert's *Dune*. The book is a classic and is really an insight piece about the dangers of politics and messianic movements. One of the characters—I think it was Duke Leto Atriedes (father of the book's protagonist, Muad'dib)—said, "Power doesn't corrupt, but it does attract people who are ultimately corruptible." That struck me as deeply important and something that seems to be playing out in U.S. politics. Of course, not all politicians are corrupt, but the temptations are strong. The only solution I see is to limit power. Absolutely limit it.

Thomas Sowell, in his book *The Vision of the Anointed*, coined the phrase "Virtue Signaling" for those people who do things because they think it makes them look good and virtuous. Appearing virtuous was important in ancient Athens. It was similarly important in the Roman Republic, but power was very constrained in both places-times. The men of Athens could vote someone out of power in a day, and they did that numerous times, including to Themistocles, victor in the Battle of Salamis and the man who saved Athenian democracy. Under the Roman Republic, a man could serve a single one-year term as Consul—and then was sent off to govern a province somewhere outside of Rome. As Rome evolved politically, the Senate, having voted Gaius Julius Caesar Dictator for Life, lost the power to vote men out of office. Augustus was, largely, a beneficent Emperor, but the same can be said for only a small minority of his successors. After the fall of the Western Empire, kings ruled in Europe, and they, like the Roman Emperors, saw no particular reason to appear virtuous—and most of them were not. Their power insulated them from the need to either be or appear to be virtuous.

Then, along comes the small group of truly remarkable men: John Adams, George Washington, James Madison, Alexander Hamilton, Thomas Jefferson, *et al.* They were familiar with the writings/musings of another group of remarkable men: Locke, Hume, Montesquieu, *et al.* That first group took the ideas of the second group and invented an entirely new form of government—the Democratic Republic. They carefully, and remarkably completely, constrained their central federal government. Their biggest failing was that they could not foresee the future, when public virtue was more to be signaled than to be lived. John Adams asserted, "Our Constitution was made only for a moral and religious people. It is wholly inadequate to the government of any other."[30]

How do moral people fare in a world in which corruptible men are attracted by power? Following the invention of the Democratic Republic by the American Founding Fathers, the global historical record becomes littered with "elections" in which the victor received 90%+ of the votes cast. These façades of democracies/republics were created by men who had ideas that they wanted to implement—wanted so badly to implement them that they were willing to rely on any means as long as they achieved their ends. Our illustrious Founding Fathers and their successors were not immune to that impulse, either—starting with the Alien and Sedition Act, efforts by President Jefferson to shutter newspapers whose editorial content opposed Mr. Jefferson, and moving through the gradual accumulation of increasing Executive Branch power peaking in the 19th century with President Lincoln's assumption of essentially dictatorial power during the Civil War.

That said, our Founding Fathers, for more than a century and a half, walked away from power after their terms of office ended (FDR excepted, but America recognized the problem with unlimited presidential terms and amended the Constitution). Nonetheless, all of the following people started off focusing on an ideal:

- Napoleon wanted to modernize the French state, and many of his reforms continue today
- Lenin wanted to create Marx's proletariat of the masses to work against economic inequality
- Mao wanted the same
- Pol Pot wanted the same
- Ho Chi Minh wanted the same, albeit under a banner of Vietnamese nationalism
- Hitler wanted to fix the world by imposing German superiority throughout Europe under German nationalism (*Deutschland Uber Alles*)
- Chavez wanted to eliminate economic inequality
- Obama wanted to realize "*The Dreams from My Father.*" Note: his father was an avowed Marxist

The problem is that human beings are not social insects programmed by evolution to march in lockstep, and lockstep is what is required for an entire nation or civilization to align with any political or social or economic model. Further, the above men deeply believed that their model was the right one—period. Since it was the right one, there was and remained no need to improve or change it, once installed. Humans don't adapt to that kind of environment well (the choice so well captured in *The Matrix* red pill/blue pill choice). Since people don't fit the model, it was and remains absolutely necessary to force those "square peg" humans into those round socio-political and economic holes. Mao said, "Political power grows from the barrel of a gun."[31]

The authors of the study *Understanding left-wing authoritarianism: Relations to the dark personality traits, altruism, and social justice commitment* (published in *Current Psychology*) talk about similarities between left-wing and right-wing authoritarians, but it is, I think, more useful to think about the political spectrum as a circle with moderates at the "northern pole" and liberalism/radicalism increasing along the left side of the circle to reach Stalin, Mao, Castro, and Che at the southern pole. Similarly, conservatism/reactionary actions increasing along the right side of the circle to reach Hitler, Mussolini, and Peron at the southern pole—right next to their radical brethren.

A major problem with governmental solutions to societal "problems" is that it's extremely hard work to identify, implement, and manage such solutions (assuming that they even exist), because governments have only two tools with which to work—money and physical coercion. The brilliance of our Founding Fathers was that they explicitly limited what their new federal government was authorized to do, and none of it was to solve the societal problems of We the People—that was up to us as individuals and as a society.

Years ago, I found myself, serendipitously, in a five-hour conversation with my airline seatmate, then Representative Xavier Becerra—he got into politics because he "wanted to help people." Not the job of government, but people like now-Secretary of HHS Becerra (and many others before him) want to do something *now*! They aren't prepared to wait and work through the sausage-making of Non-Governmental Organizations (NGO) efforts. They want to "bless" the world with their superior ideas of social perfection immediately. President Trump, similarly, has his own ideas, and he does not appear any more interested in hearing other views than do his political opponents. Like the bullet list above, many politicians today are radically impatient with the mechanics of our Constitutional Democratic Republic and would prefer to operate via executive order or let the administrative state govern rather than the slow and messy processes of three Federal Branches with their mutual, designed-in tensions.

I am a conservative, not because I find it intellectually beautiful, not because I think it provides inherently fair answers to problems, not because I think that the U.S. Constitution is perfect as it stands, and not for any of those reasons or others. I am a conservative because the principles of conservatism, as remarkably captured in a document, i.e., our Constitution—hammered out among a remarkable group of men who were both idealists and pragmatists—actually work.

The following is, very roughly, a graphic parable of how our federal system now works:[32]

Problem Solving by a Central Government

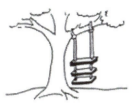

As proposed by the sponsor

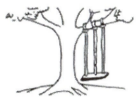

As drafted by Senate committee

As drafted by House Committee

As agreed by Joint Committee

As negotiated with the White House

What society actually needed

★ ★ ★

The Tyranny of Time

By Tom Burbage

In reality, there are three dimensions of time for most of us. The good old days, **now**, and the future.

How you view each of them shapes your life. The good old days resonate with veterans more than other groups because of the brotherhood that is founded in really challenging situations, whether surviving the Boat School (AKA, United States Naval Academy) experience, our active-duty years, or in our concern for the next generation that follows us. The Eagles, Crosby, Stills, Nash and Young, the Doors, and Woodstock all bring back fond memories—not unlike long deployments where we remember highlights and forget bad times. Our life doesn't go much farther back than that. What happened before didn't really matter; we were focused on the "Now." Some of us picked up a cudgel that our parents gave us in the firm belief that it was honorable to keep the momentum going to make America the real beacon of freedom that the world needed.

Most of the herd today has none of that in their DNA. Why is "Now" so hard to understand? Perhaps because it resides between the conflicting versions of "then" and the many versions of "when."

"Then" discussions reflect the artificial importance of how we got to where we are today. To many, it was the heroism of generations of true patriots that were willing to sacrifice all to preserve the basic freedoms of the American Constitution. To others, it was built on the back of slaves and indifference—those "many years ago" were suddenly in need of reparations. The blood and treasure expended in preserving our American way of life has somehow allowed many outsiders to object to "Now." It is not a new paradox. Abraham Lincoln and Winston Churchill both argued that we tend to forget the "loss of living history" as we distance ourselves from the last war. The "Now" generations are far removed from the sting of war. Why should we be surprised when so many allow a restatement of our history in an attempt to influence our future and many stand on the sidelines as our military is used for social experiments?

Suddenly, we seem to be tainted by the incursion of "Diversity, Equity, and Inclusion," "Critical Race Theory," and other ideologies focused on eliminating the historical references to our survival as a meritocracy. Does any of this really matter in the world we find ourselves in today? The Pilgrims, the revisionist 1619 Project, or any other reconstructionist "birthing process" is irrelevant for "Now." You can tell us that something happened in those early days, not unlike the Dinosaur age, that we all should be aware of. It's interesting, but it is irrelevant to "Now."

"When" reflects the future we see for our society, our personal lives, and the world we are going to leave behind for our kids and grandkids. The "When" may be much more difficult to define, but it is equally important. Some see a value in sustaining the vision of our Founding Fathers. But others see a very different world, where American values need to be subjugated to a Socialist ideology. There needs to be a very clear view of what that means before we, as a nation, capitulate to the other side.

How did the world change in such a short time? It changed because of *our silence*. We trusted, but we didn't verify. We trusted our schools, but we didn't verify that they were teaching and not indoctrinating. We trusted our military leaders, but we didn't verify that their incentives for promotion were not compromised. We trusted that our Service Academies were committed to developing warriors, but we didn't verify that the insidious woke culture was taking root. We were silent because we trusted. That trust is gone.

Why is "Now" so hard to understand? Perhaps because it resides between the conflicting versions of "then" and the many versions of "when." It is a classic example of the herd somehow dominating the smart intellects. The paradigm is also reflected in the basics of combat that numbers can always defeat technology if you can accept the losses. The current Ukraine situation may change that paradigm, if anyone can define "acceptable losses."

Our position as the world's leader in promoting democratic principles is clearly diminishing.

That brings us back to "Now." So . . . what do we do now? That is the question every American must face. We need to get it on the ballot, and we must ensure that the Democratic voting process can be purged of corruption. We must fight back against the false, racist statements and actions of those who are trying (and succeeding) to divide our country, our military, and our Service Academies along racial lines. We must support a return to a meritocracy-based philosophy and reject the move toward Socialism in our country.

None of these are small challenges. Our hope is that we still have enough real warriors left to accept the challenge.

★ ★ ★

Critical Race Theory in Our Military

By Randy Arrington

Critical Race Theory is Cultural Marxist Distortion Propaganda designed to give a false, perverted meaning to rewriting and misrepresenting American history and society. The goal is to create seething hatred among so-called oppressed and oppressors, brainwashing people to view everything in America from a racist perspective, making us ripe for a Communist Revolution.

Diversity is *not* a strength. Diversity without assimilation is a huge weakness for America, especially in our military ranks. In a combat foxhole, onboard a Navy submarine, strapped inside the cockpit of a fighter jet, or standing watch on the bridge of a warship, unity, assimilation, honor, respect, training, courage, and discipline are what matters most—*not* Critical Race Theory (CRT), or Diversity, Equity, and Inclusion (DEI).

In 1948, President Harry Truman ordered the United States Military to be fully desegregated. He reminded We the People of our national pledge honoring equality of opportunity and equal treatment for all American citizens as his motivation for doing so. Unfortunately, 75 years after Truman's Presidential Executive Order was implemented, un-American and racist concepts of asymmetrical treatment have crept back into the Armed Forces under the façade of so-called Critical Race Theory.

Critical Race Theory disavows the principle of equality under the law that Thomas Jefferson articulated in the Declaration of Independence and that has motivated civil-rights crusaders for several decades. It argues that this American ideal is merely a charade used by the "white privilege" majority to oppress racial factions, and as such, that America is systemically racist, to its very core. The theory argues that the only way to end perceived discrimination against racial minorities is to systematically discriminate on their behalf. In other words, Critical Race Theory states that the only remedy for past discrimination is present discrimination. And the only remedy for present discrimination is future discrimination.

This ludicrous belief in unequal treatment, prejudice, and discrimination has been wholeheartedly adopted as a newly discovered pillar of truth in many left-wing circles. Increasingly, this warped ideology is being institutionalized by numerous entities in corporate America, on university campuses, and as a plank of Democrat politics in the form of Diversity, Equity, and Inclusion Offices, and Implicit Bias Training. Regrettably, these racist concepts are now being foisted on our soldiers, sailors, and airmen.

The United States Navy released a suggested and endorsed reading list to expedite the positive "growth and development" of its sailors. Books like *White Fragility* claim that white people are inherently racist, consciously, or subconsciously, and that race is the menacing subtext for virtually all human relations. By advocating Critical Race Theory, our military is selling a poisonous ideology that rots the structure, combat readiness, and lethal fighting capacity of our Armed Forces.

As I learned during my twenty-year career as a Naval Aviator, serving alongside courageous men and women of every race and background, the military's strength is not its "diversity" but its ability to break through adversity with strong assimilation and unwavering unity of purpose. We need to train our young troops to trust, respect, fight alongside of, and, if necessary, die for their squadron mates, shipmates, and comrades on the battlefield—*not* to obsess about each other's skin pigment. Likewise, we need to instruct them to revere the Constitution that they swore to protect and defend, *not* to believe that document is a component of white supremacy in a racist conspiracy theory.

On the flight deck of an aircraft carrier at sea, I've seen several sailors confront great hardships alongside their squadron mates and watched them learn quickly to set aside their differences and concentrate on the mission at hand. These shared experiences forge formidable connections between men and women of all backgrounds, resilient bonds that are extremely rare in the civilian sector. We shouldn't sacrifice those forged relationships for the academic trends of left-wing theorists seeking university tenure.

We the People must courageously fight back against the left's effort to indoctrinate our troops with phony propaganda and turn the military into a glorified social experiment on a college campus. The military's most vital mission is to protect and defend, in combat, the Constitution and our way of life against all enemies who would destroy and vanquish our nation. Wild, far-left hypotheses of race relations are an existential threat to this mission.

Let Freedom Ring, and God Help Us.

★ ★ ★

"**Critical Race Theory in Our Military.**" This article was first published in Politics in America: Lecture Notes of a Lunatic Professor Volume 2 (IUniverse 13 December 2023)

Critical Race Theory at the U. S. Naval Academy

By Phillip Keuhlen

If you have doubts that Critical Pedagogy and Critical Race Theory indoctrination occur at the U.S. Naval Academy, then simply look at the "how to" materials on those topics supplied to faculty on the USNA Center for Teaching and Learning (CTL) website (https://www.usna.edu/CTL/Faculty_Resources/Diversity.php) and, in particular, the content of the 2019 Annual Conference on Teaching and Learning for USNA faculty found here (https://www.usna.edu/CTL/CTL_Archive/Conference2019.php).

Faculty facilitation and indoctrination in CRT and Critical Pedagogy continues in events like the CTL's 2021 Pop-up Friday sessions on avoiding "implicit bias" and "Equity at USNA."

It is telling that the Dean for CTL includes this in her bio: "She has published extensively on pedagogy and faculty development as well as in the field of British Modernism and critical theory."

It's impossible to pretend that USNA is distancing itself from CRT and wokeness when it is training its faculty to be advocates. Assurances to the contrary from the USNA Superintendent at '71's recent 50th reunion briefing are uncompelling.

But then again, why should we be concerned if those who lead and will lead the Navy and swear an oath to support and defend the Constitution are being indoctrinated with an ideology that:

"... questions the very foundations of the liberal order, including equality theory, legal reasoning, Enlightenment rationalism, and neutral principles of constitutional law"?

Remember, that is a direct quote from Richard Delgado, co-founder of CRT, on the aims of CRT, in his book *Critical Race Theory*.

★ ★ ★

Critical Race Theory at the U. S. Naval Academy, STARRS.us, 23 January 2022

Diversity, Equity, and Inclusion or Shipmate . . . You Can't Have Both!

By Brent Ramsey

According to widely accepted research, including that of the military services themselves, organizations that are effective have common values, teamwork, unity, cohesion, and a shared vision of what it takes to be successful. The literature is full of training that stresses these values. DOD, as required by Congress, conducts annual surveys to determine how things are going. The very first question asked in the 2022 survey is about cohesion! The second question asked is about connectiveness. The third question is about engagement and commitment. The fourth question is about fairness. Despite these known factors and despite the urgent problems that threaten the Navy's ability to perform its mission, what do the Navy's priorities appear to be? What we see is a leadership obsessed by race and politics that serve only to divide people into categories of skin color or gender or sexual preference. Emphasis on race and politics is a recipe for division and disunity, not the traits that lead to effective organizations. STARRS has recently published hundreds of testimonies from serving and retired airmen, soldiers, sailors, and marines expressing their opposition to the identity-politics ideology that now is widespread in the military. Those accounts can be found at STARRS. Examples abound of things the Navy does that create a lack of unity and cohesion.

- The Secretary of the Navy gave a speech recently celebrating the naming of the *USS Evans* for a Native American. The officer honored in the naming was indeed a Native American. But that had absolutely nothing to do with his heroic behavior as the Commanding Officer of a ship during WWII. He enlisted in the Navy and performed so well that he earned an appointment to the United States Naval Academy. After commissioning, he served in a variety of positions successfully up until the start of WWII, and, during the war, he was given the command of the *USS Johnston*, a Destroyer Escort. He performed exceptionally and heroically, giving his life for his country, like many brave Navy heroes before him, and his race was irrelevant to his actions. To politicize the event with the emphasis all on his being Native American and to have as the keynote speaker a Biden-administration Native American official turned an occasion for honoring an American hero into a political story for identity politics. This does a disservice to CDR Evans' memory.

- I do not oppose gays serving in the military. Many have served honorably and faithfully. However, the celebration of Pride Month is a mistake and distraction. The Navy, for the twelfth consecutive year, devoted an entire month to celebrating LGBTQI+ Pride. For Christian Sailors (or close to 70% of the Navy according to a 2019 study done by the Congressional Research Service), that is sure a fun month, inspiring unity. What one's sexual preferences are should have nothing to do with being a Sailor. To celebrate sexual orientation, including increasingly more rare proclivities that many believe are against their beliefs, is unwise and does nothing to ensure unity and a ready and lethal Navy. What does LGBTQI+ mean? L stands for lesbian. G stands for gay. B stands for bi-sexual. T stands for transgender. Q stands for queer. I stands for Intersex (whatever that means). Take note of the + at the end of the Pride initials. That + indicates that anything goes. Note the progression in this sequence. It glorifies ever smaller and outside-of-the-mainstream categories of people based on their sexual practices. How does this emphasis on what people do in private and with whom make the Navy more ready? The modern phenomenon of an overweening attention to sexual preferences and practices is never-ending and not healthy. For a Navy whose mission is outwardly focused and based on sacrifice for the good of others, this attention to and even forced celebration of individual preferences and practices is a distraction and an aggrandizement of individualism, the opposite of what the Navy should want of its Sailors. With a huge majority of the Navy being both heterosexual and Christian, to draw attention to sexual practices borders on irrational as it distracts from the mission and upsets many who are straight. You want to know why a traditional strong source of recruiting, that of Christians, is drying up . . . this is part of the reason. The Navy should discontinue celebrating "Pride." It just creates friction and dissatisfaction among many who are already serving and discourages others from joining.

- The Naval Academy is hiring a Professor of Gender and Sexuality Studies. For decades, the Navy promoted STEM (the S in STEM stands for science). Gender and Sexuality Studies hardly qualify as settled science. What does teaching highly speculative and controversial gender and sexuality theories to midshipmen have to do with creating the warriors that the Navy needs to win the nation's wars?

- Naval personnel, while on duty, dress the same in "the uniform of the day." Yet, at the Naval Academy, we find the phenomenon of 19 affinity groups whose membership is exclusive, based on arbitrary human characteristics such as skin color or ethnicity or sexual preference. If uniformity is the goal, why the promotion of cliques of people based on arbitrary characteristics that have nothing to do with qualifications or skill sets?

- The Naval Academy recently hosted its annual Diversity conference. Yet, most of those attending were minorities. Why is that? And, if the purpose of the conference was diversity, why was the attendance limited to invited guests? When a member of Calvert Group (an alumni group formed for the express purpose of supporting the USNA and its traditional values and upholding the Constitution) tried to register to attend, he was told the conference was closed and that invited guests only were allowed to attend. Why was the Diversity conference devoid of diversity of thought?

- The Naval Academy has an office of Diversity and Inclusion, consisting of five people. If diversity is a priority, why are all the people in this office Black? The Naval Academy is already extremely diverse, with hundreds of female and minority midshipmen. According to UNIVSTATS.com for the 2022–2023 class year, the enrollment at USNA was as follows:

	Total	Male	Female
American Indian/Native American	9	7	2
Asian	373	248	125
Black/African American	286	213	73
Hispanic	580	400	180
Native Hawaiian/Pacific Islander	18	13	5
Two or more races	<u>437</u>	<u>320</u>	<u>117</u>
Subtotal	1,701	1,196	509
White	2,727	1,945	782

As the figures above prove, the Naval Academy is already tremendously diverse. There is only a tiny variance between the above statistics to national racial or ethnic demographics. So, why devote the manpower and financial resources to staffing a Diversity office at the Naval Academy?

- The Naval Academy is the premier commissioning source for Naval officers and Marine Corps officers. Admission to the academy is highly selective, and USNA brags about its ranking as one of the top universities in the nation. Why are detailed selection criteria for admission to the Naval Academy a secret? As an institution paid for by tax dollars in a Constitutional Republic, doesn't the public have a right to know how the academy is operated and what admission criteria are being used to select midshipmen? Evidence that this is the situation is backed up by the fact that Students for Fair Admissions (SFFA) has filed a lawsuit against USNA, alleging that they unlawfully use the racial preferences in admissions. According to SFFA, the academy "openly admits that race is a factor" in its admissions decisions. How in good conscience can the Navy rely on racial preferences in USNA admissions in the wake of the Supreme Court ruling in the *Harvard v. SFFA* and *UNC v. SFFA* cases that ruled the use of racial preferences is a violation of law and is to be terminated? Chief Justice Roberts, in his concurrence to the *SFFA v. Harvard University* and *UNC* cases wrote, "Eliminating racial discrimination means eliminating all of it."

- There is a Navywide proliferation of Diversity staffs at every level of command all over the Navy. At every Navy website, Diversity programs are one of the first things you see. Every organization has diversity staff and programs.

- Despite evidence that the focus on DEI is detrimental to the mission, retention, and recruiting, the Navy is doubling down. For example, selection-board criteria for officer selections were just made public. Those selection criteria contain guidance emphasizing diversity and equity. The emphasis on diversity and equity undermines the concept of merit. If this criterion results in less-qualified members being promoted to higher grades, the Navy will see the results in terms of declining readiness, morale, retention, and war-fighting effectiveness. Will future leaders selected based on diversity and equity be able to help win the nation's wars?

For the Navy to be successful, teamwork and cohesion must be paramount. A unique concept has been at work for centuries, and that concept is that of *shipmates*. The concept is explored in depth in my article "Shipmates: A Dying Breed." In a ship at sea, going into harm's way impacts every Sailor in the exact same way. If the ship sinks, everyone's life is at risk. That is why in bootcamp, Sailors are trained the way they are . . . to be part of a team. Everyone is treated the same . . . the same uniform, the same berthing accommodations, the same chow, the same training. You are trained to have your shipmate's back and he or she has yours. Your race, your ethnicity, your gender, and your sexual preference don't mean a thing. When aboard ship, you will be a cog in a vast and complicated machine, and every cog needs to concentrate on its part, its duty to keep the ship safe and functioning. Individualism goes out the window. The only thing important is to be an effective part of the team to keep the ship safe and to allow the ship to complete its mission. Unity of thought and behavior is paramount. The needs of the ship and safe navigation come first—or lives are threatened. For the Navy to emphasize race and ethnicity through its Diversity and Equity ideology is a big mistake, as it erodes the concept of a *shipmate*. *This article appeared in Armed Forces Press.*

★ ★ ★

Diversity, Equity, and Inclusion or Shipmate . . . You Can't Have Both! Armed Forces Press, 3 December 2023

That Damn Pandemic and the Ultimate School Board

By Tom Burbage

In the summer of 2020, in the height of a global pandemic, an assault on the United States started in a number of cities across the country. Burning, looting, murders, often without lawful intervention, were common threads. But those places were "somewhere else," not here, where we lived. Unless you lived there.

Here, where we lived, our kids were suffering through a year of virtual confinement and mask wearing as an alternative to the school environment we grew up in. A new educational system evolved with virtual classrooms. Parents, also confined to home and the "virtual business environment" suddenly got a glimpse of what was actually being taught. For years, they had gone about their work focus and assumed teaching was still "like we remembered it." It was a rude awakening and an awareness that may never have happened were it not for "that damn pandemic."

Looking over the shoulder of the "kids in the kitchen attending class" was traumatic—even more so when teachers appealed to prevent parents from watching. Digging into the details revealed many curious but buried facts in the evolution of school curriculum as early as elementary school. Suddenly, parents were able to experience the content of much of the new educational curriculum, centered on critical race theory, being broadcast into their home. This awareness resulted in the "revolt" against school boards we are seeing today. The proponents of indoctrinating the next generation of our society with a race-based oppressor-and-oppressed ideology and a distorted view of our history became a target for enraged parents. In the parents' view, school boards and teachers should be trusted wards of our children, not advocates for radical change. The momentum of this movement would not likely have been challenged were it not for "that damn pandemic."

So, what does this have to do with our military?

It comes down to two fundamental societal elements: families and tribes. The formal definition of family is that it is the basic unit of society, traditionally consisting of two parents rearing their children. The movement to eliminate the nuclear family had been underway since the Great Society initiative of the Johnson Administration that incentivized fatherless families among the poor and underprivileged. The roots of Marxism began growing and accelerated with the opportunism of the BLM movement. The group cleverly leveraged the emotional branding of the phrase to gain support among the clueless and uneducated but openly stated an objective to further eliminate the nuclear family as a key step to achieving their educational objectives. A tribe is a social group composed of numerous families, or generations having a shared ancestry and language. The military family is in reality a very large tribe of people that includes several generations that have fought and died to defend our Constitution and our American way of life. These elements are powerful forces in controlling the pendulum swings that either drive us toward an unacceptable future or bring us back to a sense of stability. Like the parental awakening to the danger of the Socialist/Marxist agenda infiltrating our school systems, there is a new awakening that the same cancer is infiltrating our military.

Like the local school board when facing an outraged parents group, fireworks ignite. Our military veterans mirror the parent community, concerned that their offspring, the next generation of military leaders, are now experiencing the same infiltration as our school systems.

The line between studying ideologies that are diametrically opposed to our Constitution to understand evolving threats to our way of life and indoctrinating our youth and military inductees into that ideology is subtle and has dramatically different objectives. Our military leadership has always been the world standard in developing the warrior culture. It is now the "Ultimate School Board," trying to respond to the enraged-parent community that is committed to taking back our unique American military culture. Parents entrust their children to teachers and schools and have expectations for the experiences of the education process. When they find the process has taken a dark turn, they will react, although sometimes too late to be effective. In a very similar way, when political military leadership takes a similar dark turn, the veteran/alumni group will also stand up and be counted.

There has been a withering rebuttal of our military leadership in the last few months, but, in reality, it goes back several years. There is sand in the gears of our well-oiled military machine. It comes from two directions—the school-board agenda pushing social experimentation into the military and the "Parental" objection to dilution of the warrior culture so important to maintaining our leadership on the world stage. My money's on the parents.

★ ★ ★

Politics and the Academy Boards of Visitors

By Brent Ramsey

10 US Code §8468. Board of Visitors
(a) A Board of Visitors to the Naval Academy is constituted annually of:
(1) the chairman of the Committee on Armed Services of the Senate, or his designee;
(2) three other members of the Senate designated by the Vice President or the President pro tempore of the Senate, two of whom are members of the Committee on Appropriations of the Senate;
(3) the chairman of the Committee on Armed Services of the House of Representatives, or his designee;
(4) four other members of the House of Representatives designated by the Speaker of the House of Representatives, two of whom are members of the Committee on Appropriations of the House of Representatives; and
(5) six persons designated by the President.
(b) The persons designated by the President serve for three years each, except that any member whose term of office has expired shall continue to serve until his successor is appointed.

The public law cited above establishes the Board of Visitors (BOV) for the United States Naval Academy. Similar provisions of law establish similar boards for the United States Military Academy and the United States Air Force Academy. These BOVs are oversight boards, created by Congress to investigate, oversee, and make written recommendations about the

Academies to the United States House Armed Services Committee, the United States Senate Armed Services Committee and/or the President of the United States. The duties of the BOVs include the following:

The Board shall inquire into the state of morale and discipline, the curriculum, instruction, physical equipment, fiscal affairs, academic methods, and other matters relating to the Academy that the Board decides to consider.

On 2 February 2021, Secretary of Defense Lloyd Austin suspended the three Academy BOVs, and those boards were inactivated on that date and unable to perform their Congressionally mandated duties. On 8 September 2021, President Biden demanded all the President Trump-appointed members of the BOVs, including Heidi Stirrup, a member of the USAFA BOV, resign by the close of business that day or be terminated as a member of the BOV at 6 p.m. that day.

On 15 July 2021, Heidi Stirrup filed a lawsuit in federal court challenging the suspension of the USAFA BOV. On 17 August 2021, additional plaintiffs were added to that lawsuit, including Congressmen Mark Green and Ralph Norman, four cadets and five of their parents, to also challenge the suspension of the BOVs at West Point and the Naval Academy. On 8 September 2021, Sean Spicer, former White House Press Secretary for President Trump, became a Plaintiff as a member of the Naval Academy BOV. As of October 20, 2023, that case is still in process. Subsequent to the lawsuit being filed, the Biden Administration made new appointments to the BOVs, and they resumed functioning with Biden appointees performing the duties.

The Naval Academy Board of Visitors has been in existence since 1879. Dozens of Congresses over a 140+ year time span have supported their missions and functions as a necessary oversight body to provide advice and counsel on the well-being of Naval Academy midshipmen. Yet, in 2021, the current President decided to unilaterally ignore the Congressional intent and political balance that spans much of the nation's history to inject political imbalance into the oversight of the military academies. When the press inquired as to why such an unprecedented firing of 18 Trump appointees had occurred so abruptly and without any advance warning, the President's press secretary answered that the President needed members on the boards that were "qualified to serve" and represented "our values." Apparently, General Jack Keane, a former Vice Chief of Staff of the Army, and Lt. Gen.

H.R. McMaster, the former National Security Advisor, were among those the President questioned as being qualified to serve and whose values are somehow not compatible with his.

As a matter of law and prudence, these actions against legally appointed and serving BOV members are troubling. DoD's suspension of the academy boards has shut them down completely, thereby preventing any BOV oversight or recommendations since February 2021, at a time when each of the academies has unprecedented problems, including honor scandals, COVID, resistance to mandatory vaccinations, critical race theory teachings, purging of "extremists" defined only as "white supremacists" but not including Antifa or BLM, sexual assault, and suicides. It is hard to fathom how DoD legally and morally can justify so blatantly eliminating Congress's oversight of the academies through its BOVs, especially when Article I, Section 8, Clause 14 of the U.S. Constitution provides that "[The Congress shall have the Power . . .] To make Rules for the Government and Regulation of the land and naval Forces."

Without a doubt, appropriate legal authorities within the administration are aware of court rulings on the matter of BOV tenure. A recent Supreme Court decision reaffirms, citing two previous Supreme Court decisions, that the U.S. Constitution allows "tenure protection," meaning that Congress can prevent the President from replacing lower-level "inferior officers" appointed by the President who, like members of the BOVs, make recommendations but have little or no authority. Congress passed statutes setting the terms of BOV members. The threshold issue is whether by passing these statutes setting these terms, Congress intended to prohibit the President from removing a Presidential appointee prior to the expiration of the term set by statute. According to the Supreme Court, the President cannot fire and replace BOV appointees before the expiration of their statutory terms if Congress intended to prohibit early termination of Presidential appointments to the BOV. Not allowing these BOV members lawfully appointed by the previous President to serve out their terms set by statute smacks of political gamesmanship, and those who put the President up to it should be identified and rooted out of government. No President—of any party—should put up with such nonsense. Not only are the lives and reputations of 18 distinguished Americans needlessly sullied and tarnished, but the important role they have been playing for years to help and attend to the needs of our academies' cadets and midshipmen has been eliminated, replaced with confusion and acrimony. The cadets at the U.S. Military Academy must be especially demoralized and confused, as the publicly maligned Lt. General H.R. McMaster was presented with the Distinguished West Point graduate award on 11 September, 2021.

This attempted removal of lawfully appointed BOV members is another example of politics invading the US military. Politics has no place in the military. The military must be free of all politics in order to do its job of defending the nation. If military members are swayed by the ebb and flow of politics, election to election, it will irreparably damage its effectiveness as a warfighting organization. The military should have a single-minded focus on supporting and defending the Constitution of the United States and nothing else.

A copy of Heidi Stirrup's answer to the request that she resign is quoted in full below:

"September 12, 2021

Catherine M. Russell Assistant to the President Director, White House Office of Presidential Personnel

Dear Ms. Russell:

I am responding to your email and letter to me dated September 8, 2021, requesting my resignation as a Member of the Air Force Academy Board of Visitors, and stating that if you have not received that resignation by the close of business on that date, my position with the Board will be terminated effective 6 p.m. on that date. Please be advised that I have not resigned and will not do so. Moreover, I have received no official notification that my position with the Board has been terminated, and therefore consider that it has not been terminated. I understand that the resignations of other academy Board of Visitor members have also been requested and termination of their Board positions also threatened. These resignation demands and threatened terminations are unprecedented and, I believe, bad policy because they are divisive, destroy the politically balanced structure of the Boards of Visitors as created by Congress, and deprive the academies of the diversity of experience, knowledge, and perspectives the academies traditionally have had and need in order for the academies to fulfill their missions competently and in a manner reflective of all of the American people. Moreover, and most importantly, termination of membership of an academy Board of Visitors would be illegal. Congressional statutes expressly provide the dates of expiration for academy Boards of Visitors membership positions. For example, my tenure as a Member of the Air Force Academy Board of Visitors is governed by 10 U.S.C. § 9455(b), (c), which when applied requires that my tenure does not expire until after December 30, 2021. This result is supported by a recent United States Supreme Court decision reaffirming that Congress can continue to provide "tenure protection," recognized for decades as valid by the courts, for Presidential appointments of inferior officers with narrowly defined duties, which would include Members of academy Boards of Visitors, who have no authority but can only make recommendations and issue reports as narrowly defined by statute. See *Seila Law LLC v. Consumer Fin. Prot. Bureau*, 140 S. Ct. 2183, 2192, citing *United States v. Perkins*, 116 U. S. 483, 6 S. Ct. 449, 29 L. Ed. 700, 21 Ct. Cl. 499 (1886); *Morrison v. Olson*, 487 U. S. 654, 108 S. Ct. 2597, 101 L. Ed. 2d 569 (1988). Please identify to me any legal authority believed to authorize the termination of my Membership to the Air Force Academy Board of Visitors. *This article appeared in CD Media.*

★ ★ ★

Politics and the Academy Boards of Visitors, Creative Destruction Media, 19 September 2021

A Monster at the Door

By Bruce Davey

It is worth contemplating that CRT is basically a frontal attack on the *entire* nation, its foundations, its historical record, and its heroes, whereas mixing up of genders, focus on plants and animals above humans, and even "reproductive rights" are just tickling around the edges with pernicious influence but lacking a method or manner to break open the society as a whole. They attract "nuisance groups" who conduct gay parades, climb and live in old-growth timber, or kill off their young, but are no real threat to our precious nation. Given that their spiritual underpinning is clearly lacking and their keels are crooked, they will certainly bend in time and find themselves hard pressed on a lee shore. The philosophy and engineering behind CRT are much more gainfully structured. The language is bent in confusing and disruptive fashion; the slipping of purposely conceived ulterior destinations into what are seen as worthwhile goals, and then inserting a "Get Well" card into a wrapping that is by its very nature conflict-inducing, cohesiveness-destroying, and, ultimately, fatal to the host, should be terrifying to anyone who loves these United States.

Forget gender issues, vaccination edicts, and the donning of face masks—we can fight them on our own. We must not be distracted by mosquito bites or bee stings when there is a monster at the door.

★ ★ ★

The Navy and Diversity

By Brent Ramsey

In recent years, the Navy has put a lot of emphasis on "Diversity." If you visit the Navy's recruiting website, www.navy.com, on the first page, it has an entry, "Who we are." When you access that page, it contains three entries:

- Women in the Navy: With a picture of five women Sailors at sea.
- Diversity and Equity: With a picture of a Black female Sailor in the foreground.
- Reserve: With a picture of a Black male Sailor looking out over the sea.

Under "THE NAVY'S COMMITMENT TO DIVERSITY & EQUITY" it says, "Why are diversity and equity important to the Navy?" followed by this statement: "We believe that leveraging our diversity is the key to reaching the Navy's peak potential, both as a workplace and as a defense force. We also believe that when leaders tap into the energy and capability of an actively inclusive team, we achieve top performance. We know that different perspectives shine light into our blind spots, illuminating what we wouldn't otherwise see." No evidence is provided proving diversity has any connection with warfighting success. The statement is one of wishful thinking. The fact is that there is a dearth of evidence that diversity has anything to do with warfighting effectiveness and lethality. If readers know of such evidence, I would love to hear about it.

The Navy has a diversity policy statement. It says: "The Department of the Navy (DON) is fully committed to creating and maintaining an environment which supports Diversity and Inclusion (D&I). The DON understands that its strength is dependent upon its people and the different perspectives, talents, and abilities they bring to the workplace. Prevailing today and adapting to the emerging security environment of tomorrow necessitate the continued attraction of our nation's increasingly inclusive and diverse workforce. The DON is committed to the inclusion of our diverse collection of Sailors, Marines, and civilians into all aspects of the organization's operations. Successfully including all personnel creates an environment that motivates innovation and provides fresh perspectives, which allows the DON to reach its maximum warfighting potential. Practicing D&I principles allows the DON to fully leverage the wealth of knowledge, experience, and perspectives from all of our people. The DON's core values—Honor, Courage, and Commitment—reinforce our promise to respect others and provide equal opportunity for all people. Our commitment to each other is second to none, and we must match our dedication to an inclusive environment that assures the success of our core mission. Therefore, I ask every Sailor, Marine, and civilian to join me in ensuring our workforce actively includes all perspectives in order to harness the powerful benefits of D&I." Again, no evidence is presented that a diverse force is more effective, more lethal to our enemies. This is an example of pandering to the political left with no substance about how to achieve the goal of more diversity or any proof that attaining that diversity makes the Navy more dangerous to our enemies.

The Navy stood up an organization, Task Force One Navy (TF1N), in July 2020 with great fanfare to examine diversity and how to achieve it. At its creation, the Navy had 8% Black officers.[33] The 2022 DOD Military Demographics Report shows the Navy now at 7.9% Black officers. The TF1N report had dozens of recommendations about how the Navy would improve the number of minority officers. The only conclusion that can be reached by this outcome after more than 2 years is that the Navy is clueless about what it takes to attract more minorities to the Navy. Maybe enticing more minorities to the Navy is more than clichés about diversity and inclusion.

At the United States Naval Academy (USNA), the premier educational institution for training career Naval officers, there is an Office of Diversity,

Equity, and Inclusion headed by a CAPT with a staff of five, all Black. What possible purpose could this office serve? Those selected for USNA are already there, selected by the admissions staff, based on the criteria established by the Navy for attendance. The Navy's thumb is already on the admissions scale in favor of two elite groups coveted by the Academy for two entirely different reasons. The first group are elite athletes who get in for their athletic prowess, not their test scores or GPA. This group already has a high percentage of minorities, a necessity to attempt to be successful in the elite sports like football and basketball. This phenomenon of the USNA competing in Division I sports is entirely superfluous to training naval officers and only serves to pander to the alumni who desire such prestige sports to be provided by their alma mater. The second group are certain minorities who get in because the standards for Black and Hispanic minorities are lower than the standards for whites and Asians. We know for a fact that standards for certain minorities are lower, from the statements made by the U.S. Solicitor General's statement to the Supreme Court arguing in favor of racial preferences for the military in the *SFFA v. Harvard* and *SFFA v. UNC* cases. Solicitor General Elizabeth B. Prelogar told the justices on Oct. 31, 2022, "So, it is a critical national security imperative to attain diversity within the officer corps. And, at present, it's not possible to achieve that diversity without race-conscious admissions, including at the nation's Service Academies." She offered this statement without proof that diversity adds to the lethality or effectiveness of our nation's fighting forces.

The impression left of these glimpses of the Navy described above is one of pandering, of painting a false image of the Navy to attempt to appeal to women and minorities to join, to convince the public and potential recruits that the Navy is politically correct.

Part 2

The dictionary defines "diversity" as, "the state or fact of being diverse: difference: unlikeness." Another definition—and the one the Navy is presumably alluding to—is: "the inclusion of individuals representing more than one national origin, color, religion, socioeconomic stratum, sexual orientation, etc."

By policy and recruiting emphasis, the Navy tells you exactly what diversity they want. First and foremost, they want skin-color diversity. How do we know this aside from the pictures at the recruiting website? Because Task Force One Navy (TF1N) was all about attracting more Blacks, Hispanics, and native Americans to Navy *officer* ranks only. The report documents clearly that the Navy is already overrepresented in enlisted minorities.[34] Thus, its laser focus is promotion of the idea that the Navy needs more Black and Hispanic officers, although the preeminent emphasis seems to be for more Black officers. Why this conclusion? On Admiral Gilday's CNO watch, he urged all hands to read *How to Be an Anti-Racist* by Ibram X. Kendi. Kendi is an avowed racist, based on his own words. He openly advocates for racism against Whites and for the supremacy of Blacks. Kendi is an advocate for race-based discrimination, arguing "the only remedy to past discrimination is present discrimination."[35] Increasing the number of Black officers apparently became the priority under ADM Gilday. After the George Floyd tragedy, the Navy openly embraced support for Black Lives Matter.[36]

In the latest DOD Military Demographics Report 2021, the Navy officer corps is still only 7.9% Black versus the goal of 13% established by the TF1N report. Since the TF1N report, the Navy's numbers still have not gone up. The Army reports 12% of its officer corps is Black in that same report, so the Navy's result seems unusual. Overall, the 2021 report shows 13,361 minority officers in the Navy. One would have to conclude that there are no meaningful barriers to qualified minorities to join the officer corps of the services, since more than 54,000 minority officers are currently serving.[37] According to reports of the Department of Education, there were more than 200,000 graduates with either Bachelor's or Master's degrees in school year 2020–2021.[38] With qualifications easily met by thousands of minorities each year, why is it that fewer Blacks and Hispanics choose to serve? There is zero credible evidence of barriers to service. Quite the opposite is the case, with the services bending over backwards to try to recruit minorities, especially with programs that reach out to the Historically Black Colleges and Universities (HBCUs) and Minority Serving Institutions (MSIs) to try to attract minorities to the services. The TF1N report contains 63 references to HBCUs and MSIs.[39] It is plain that the Navy claiming to want diversity is not convincing to most of those eligible to serve and that the Navy lags behind the Army substantially in providing a convincing argument on why serving the nation is a good idea.

The other diversities being sought are in sexual practice and gender diversity. This comes through loud and clear at the website, in recruiting media and images, and through such phenomenon of the military's celebration of Pride Month and, lately, even promotion of drag-queen shows onboard ships and a drag queen being used for recruiting (which the Navy has now retreated from due to negative reaction from the public). Add to that the nonsensical promotion of popular progressive practice of what pronouns to use, and the farcical and unserious nature of our current leadership becomes clear. Just days ago, DOD had to retreat when it was pointed out its new pronoun policy for awards was both ungrammatical and unpopular with the public. Even our outgoing Chairman of the Joint Chiefs of Staff objected.

If diversity is such a priority, why pick out only Blacks, gays, and transgenders to emphasize? The definition above includes the categories of national origin, religion, and socioeconomic stratum. Why no commercials to get Christians or Muslims to join? Why no ads to attract immigrants? Why are there no recruiters in the inner cities among high concentrations of young socio-disadvantaged minorities? The answer is plain that the diversity push is motivated by the popular canard promoted by the left that the U.S. is still a racist nation and must provide racial preferences to minorities. The recent Supreme Court cases *Harvard v. Students for Fair Admissions* and *UNC v. Students for Fair Admissions* emphatically ruled against using race in college admissions as violations of equal-protection law.

Part 3

The truth is that the racial- and sexual-identity politics advocates have gained control over our nation's culture, and this has seeped into our military, where it does not belong.

Although we hear a lot about systemic racism and White supremacy, there is no evidence that such is present in today's military. DOD's own internal review shows that fewer than 2% of serving DOD personnel identify racism as a problem in DOD.[40] DOD's extremism stand-down and subsequent reporting revealed fewer than 100 incidents of extremism in the last reporting year out of a DOD force of more than 2.1 million, an astonishingly tiny number.[41]

So, why all the fuss over diversity, diversity, diversity, when the services are already substantially diverse and have few internal complaints of racism or extremist activity? Could it be purely a conformity reflex and a reflection of the hold the left has over our military leadership, which is supposed to be apolitical?

Nowhere at the Navy site is proof offered that diversity is essential for combat effectiveness. We do have a lot of counter evidence that diversity advocacy is flawed or counter to fielding a lethal military, such as:

- Col (Ret) Bill Prince, U.S. Army Special Forces with 11 combat deployments, attended the recent USMA Diversity conference. In his recent article, he quotes the USMA's Chief Data Officer, Col. Paul F. Evangelista '96, who, in commenting on attempts to measure the effectiveness of DEI, said, "**We don't have the data.**"[42] West Point's Chief Data Officer answered the DEI question candidly: "**We don't have the data.**"

- BG (Ret) Ernie Audino, U.S. Army, nails the issue precisely in his article saying,[43] "Because, if Prelogar and those generals are right, i.e., that racial diversity in our officer corps is a "national security imperative," then the services would at least track racial percentages in their mandatory assessments of unit combat readiness, **but they don't. Racial diversity is not included, and never has been.**"

- CDR (Ret) Phil Keuhlen, USN, is a former Commanding Officer of a nuclear-powered attack submarine. He conducted detailed analysis of the TF1N project and its two sources cited to prove better performance due to diversity. His conclusion . . . neither source used by TF1N passes muster. His detailed analysis can be found in a succeeding essay and at this noted link.[44]

- COL (Ret) Bing West, USMC, is one of the most decorated combat veterans in our nation's history. COL West also served as Assistant

Secretary of Defense for International Security Affairs under President Reagan. Bing's article "The Military's Perilous Experiment" ought to give our military's leaders pause in their headlong pursuit of diversity. He writes, "Inside the military, however, another criterion has taken central booking: diversity. The focus has shifted toward emphasizing gender and racial equality, particularly in leadership positions. Diversity has replaced lethality as the lodestone for the military." "It's all about warfighting readiness," Chief of Naval Personnel Vice Admiral John Nowell Jr. said. "We know that diverse teams that are led inclusively will perform better." On one level, that sentence is a tautology; every individual is unique and, therefore, every team is diverse. On another level, the admiral is speaking in code. He is implying that the services have been underperforming because they have not properly rewarded diversity. As a Marine veteran, I find this disconcerting. From boot training on, Marines are taught to put aside diversity, not to emphasize it." The entire article can be found at this link.[45]

If the Navy were truly interested in the merits of diversity, there is ample study on the subject by eminent scholars, many of them Black. The lack of intellectual curiosity on the topic of diversity by Navy leadership is astounding and an apparent manifestation of the politicization of the senior officer corps to the detriment of our core mission of preparing to fight and win our nation's wars. *This article appears in Real Clear Defense.*

★ ★ ★

The Navy and Diversity, Real Clear Defense, 7 October 2023

Dinner with Joe

By Bruce Davey

At 1800—6 p.m. to the landlubbers among us—every night when my father was not at sea, dinner was served in our home by my sainted mother. Our only requirement was to wash our hands. Often, this was accomplished by running water over the filth and then "drying" the dirt onto a clean towel. The end result was a superficial accomplishment of present targets and an unfortunate increase of unpleasantness in the near future. It would not be long before the "clean hands, pure heart" fiction was discovered and the consequences of the supposed good work were revealed.

When one contemplates the actions of the present administration, the parallels abound. A deep and abiding feeling of communion with the underprivileged and downtrodden that was to be manifest in the new President's support of BLM and Antifa resulted in the filthy towels of buildings burning and murder in the streets. The hope that a 9-11-21 ending of the Afghan conflict would polish the foreign-affairs record of President Biden had an irresistible appeal to those who deal only in spin and optics. It did not take long for the dirty towel again to become apparent as we watched, horrified, as civilians dropped from aircraft underbellies and our valiant service members were murdered.

The historical record of dalliance with superficialities and veneer continued with a wafer-thin belief structure, which led to cancellation of pipelines, disastrously limiting drilling, and overseeing the dismemberment of America's self-sufficiency in energy production. The good look that was presented to his political allies was soon revealed to be a drive into subservience to Middle Eastern cartels and Russian petroleum oligarchs. The hands looked sort of nice at the table, but filth reigned everywhere after dinner.

Questions now abound with regard to the future continuance of this perplexing pattern. Clearly, the ongoing diminution of our Armed Forces on the twin altars of vaccine mandates and the poison of CRT/DEI, although placating and pleasing the "woke" anti-Americans, and presenting the President, the CJCS, and the Secretaries of each Armed Service with nice clean hands at the dinner table, hide a breakdown of the cohesiveness and effectiveness of our Armed Forces. The fabric of our defense is sullied and dangerously thinned.

With only a glance slightly further into the future, we see the pattern beginning to be outlined again in our negotiations with Iran. In what is clear to the Mullahs, Joe Biden is again looking for a pretty picture to present to his followers displaying his skill and cunning in international relations. Our enemies in Iran recognize the huge ego and diminished cognitive capability which combine to make our President an easy mark. They will give him some clean hands to hold up to the light while promising "Peace in our time." The adoring press and a nation of sheep will fail to notice the ruined towels that lie in a pile beside the nuclear codes.

★ ★ ★

What Diversity, Equity, and Inclusion Is Doing to the Navy

By Brent Ramsey

At the U.S. Navy website, it says:

"Navy focuses on DEI in three ways:
For Diversity, Navy measures how individual communities compare to the Department of Labor comparable-civilian equivalent, officer, and enlisted demographics along race, gender, and ethnic lines. This ensures there are no unintended barriers to entry and helps focus Navy recruiting efforts to bring in the right available talent."

What are Navy's demographics? The Navy matches quite closely the national metrics. The Navy falls short only in female officers. The female-officers deficit is not due to any barrier to entry. It is due to females largely being uninterested in military service. How does this comparison ensure "there are no unintended barriers to entry"? The standards for admission to the Navy are the same for everyone except that Blacks and Hispanics are offered preferences based on skin color. That unfair advantage is what the Supreme Court ruled against by a 6–3 majority in the *Students for Fair Admissions vs Harvard* and *UNC* cases. Is Navy ignoring the Supreme Court ruling that racial preferences are illegal?

Navy then says:

For Equity, Navy looks at key billets, along with detailing, advancement/promotion statistics to ensure every Sailor has the same opportunity for professional growth and development. This enhances organizational loyalty, encouraging Sailors toward a Navy career because they can see themselves in senior Navy leaders, both officer and enlisted."

The Navy uses the word "equity" but actually says "every Sailor has the same opportunity," i.e., equal opportunity. Equal opportunity is not equity. Equity is equal results. Using "equal opportunity" enhances organizational loyalty. Using "equity" would create dissension in the ranks. This is an example of the Navy wanting to have it both ways. It appears they are using the word "equity" but practicing "equal opportunity."

Navy then says:

"For Inclusion, Navy uses a variety of surveys to assess whether or not its workforce feels included and connected to mission and leaders at all levels. This reflects human psychology as it relates to teambuilding, where personnel who feel excluded and disconnected are more likely to both underperform and conduct destructive behaviors."

Assessing workforce welfare? Why doesn't the Navy report what DOD climate assessments reveal? Those assessments include Navy personnel. Those surveys show fewer than 2% of the workforce indicate that racism is a problem. With such a low percentage, one wonders: Why all the emphasis on DEI? I can only speculate that it is a political calculation by Navy leadership.

You will find similar, vague DEI statements at virtually every Navy site. It is a ubiquitous and cloying obeisance without substance to political masters that is sucking up millions of precious dollars and manpower, detracting from the core mission, and harming morale and focus on readiness. Both internal and external reports document actual problems that the Navy should be dealing with ... horrific maintenance backlogs, overstressed crews due to longer and longer deployments, low readiness rates documented by internal Navy reports, the GAO, the Heritage Foundation, and the plague of suicide in the ranks.

An emphasis on DEI is the exact opposite of the core of what it is to be a Navy, a military service. For example, all Navy personnel are required to wear a uniform while on duty. The word "uniform" means: "Identical or consistent. Without variation in detail." Yet the Navy is now laser focused on diversity, which is the opposite of uniformity. Isn't focusing on superficial attributes such as the color of your skin or ethnicity a recipe for creating division, not uniformity?

All the attention to DEI has become an obsession and has been institutionalized. It is a system based on a belief with no actual evidence that diversity improves performance or readiness. We know this is so because diversity advocates operate behind closed doors and will not allow anyone in who wants to ask questions regarding this emphasis and what the emphasis is doing to our readiness. Navy leadership preaches DEI without supplying any evidence that DEI is objectively making the Navy better, stronger, and more ready. The opposite is true. At the recent USMA Diversity conference, the West Point Data Officer confirmed that the Army has no data supporting the claim that diversity improves unit performance or readiness. Furthermore, there is increasingly more evidence that the promotion of DEI is having the exact opposite effect from that intended. Across the nation, there is tremendous pushback against DEI in academia, in the corporate world, and in state government. How can Navy leadership continue to ignore what is going on all around them?

The Navy has approximately 230 flag officers. They are the leaders of our Navy. These are the best and brightest. These officers were bred and trained based on our core values of honor, courage, and commitment. I respectfully challenge these officers to objectively evaluate what the Navy is achieving with its focus on DEI instead of our traditional values of honor, courage, and commitment. That is the Navy I served in for 30 years, the Navy that helped topple the Soviet Union. The Navy must reject DEI. Focus on diversity can only divide. Focus on equity undermines merit. Focus on inclusion is a fool's errand, as personal neediness breeds weak minds and weak spirits. The Navy must go back to basics and renew our commitment to our core values and concentrate with laser focus on getting the Navy ready to fight and defend our nation. Instead of continuing to promote DEI and other political topics, the Navy's leaders should concentrate on solving the serious problems facing the Navy, those challenges posed by the People's Republic of China, Russia, Iran, the DPRK, and radical Islam. Concentrate on building a Navy that citizens want to join. Take care of those who do join, and focus on the mission. And finally, focus on readiness above all else for our ships and aircraft, so that we will be able to execute our sacred mission to defend the nation against our enemies.

This article originally appeared at AmericanThinker.com[46]

★ ★ ★

How Social Justice Is Killing the Military

By Brent Ramsey and Michael D. Pefley

"Some 45% of the military identify as members of minority groups," the Department of Defense (DoD) reported in its 2021 Demographics Report.[47] That number is higher than the 42% of the U.S. population who identify as non-white.[48] "We all bleed RED," service members say. Years ago, when a commanding general asked for a report on all disciplinary cases by color, a Black Chief Master Sergeant from rural Alabama whose parents grew up under actual racism, reported back, "All our soldiers with disciplinary cases are green."

The least racist people in America are in our military. It is obvious to all who have served because one's identity or race is irrelevant when people are trying to kill you and your teammates. No one cares about race when bullets are whizzing past your head.

So why has President Biden declared White supremacy as America's number-one threat? Why has the Secretary of Defense created a Counter Extremism Working Group to uncover extremism in the ranks? The actual record of racial/radical incidents is not the reason—the number of cases is tiny. With 2.4 million in uniform, DoD reports fewer than 100 annual incidents of "extremism," without defining "extremism." When the DoD spokesman was asked for specifics of extremism, he did not have them. If the government declared that "Trolls" are a threat to the Republic and began an investigation, would we nod our heads and say, "I want to know about Troll Rage" because as a child we learned from fairy tales how vicious trolls can be? Now the fairy tales are about White racism.

Yet, while admitting that 45 percent of the military identified as minorities, the DoD insisted that diversity, equity, and inclusion were military necessities.[49] Diversity, Equity, and Inclusion (DEI) is the language of Critical Race Theory (CRT). Diversity is the promotion of division, based on identity quotas or targets. Inclusion is code for exclusion of specific groups based on race, gender, and sexual identity. *Equity* is CRT's ambiguous terminology, easily confused with *equality*. Equity is equality of outcomes (usually quotas), which is reformulated Marxism.

Stand Together Against Racism and Radicalism in the Services, Inc. ("STARRS") was formed in 2021 to educate the services and America that our once-capable military has been diverted from its core national defense mission into social-justice warriorship. STARRS has incontrovertible evidence that this divisive threat spans all branches.[50]

DEI Torpedoes Navy Readiness
With DEI implemented throughout the Navy, a return to selection-board photographs is proposed to increase equitable minority promotion (quotas). Ibram X. Kendi, author of *How to Be an Anti-Racist*, has lectured midshipmen that we need racism against "Whites" to "make up for past racism."[51] Kendi's book *How to Be an Anti-Racist* was on the Chief of Naval Operations' recommended reading list, and the criminal-justice system is bogusly defined as inherently racist and the new Jim Crow. A report commissioned by Senator Tom Cotton and representatives Jim Banks, Mike Gallagher, and Dan Crenshaw quoted a Black officer as saying: "Sometimes I think we care more about whether we have enough diversity officers than if we'll survive a fight with the Chinese."[52] The report also revealed that readiness problems are due, in part, to diversity emphasis taking priority over operational training requirements. That's why ships have collided and arson has destroyed a capital ship, with a replacement cost of more than $3.4B. Ship handling and firefighting appear to be less important than DEI.

Air/Space Forces Shoot Down Core Values
The United States Air Force Academy (USAFA) has not been responsive to 15 Freedom of Information Act requests submitted by STARRS, including a

summary of "racially based incidents" that justified a DEI push for cadets. If there's nothing to hide, then why no response? One of the Space Force's top-rated lieutenant colonels, Matthew Lohmeier, was removed from command and separated without benefits because he wrote and spoke the truth about racism/radicalism in the Space Force. His concerns about divisiveness of the Marxist push were dismissed and labelled "political," despite his book being pre-approved. At USAFA, a separate cadet-diversity chain of command has been established outside of the regular squadron chain, which is akin to the Communist Party's political officers within Russian military units.

Diversity Blows Up West (Woke) Point
The Chairman of Joint Chiefs of Staff defended the CRT program and testified to Congress that he wanted to understand "White Rage."[53] The United States Military Academy has been teaching CRT along with White-rage seminars as reimagined Shakespeare. Ta-Nehisi Coates spent two days at West Point in 2018 speaking about race. Coates's toxic views, published in *The Atlantic*[54] in 2014, hold that all Whites are racist and calls for trillions in reparations for all Blacks. According to a spokesman for Congressman Crenshaw's office, more than 500 whistleblower complaints have been submitted about conditions in the services. If 500 have already come forward, how many thousands have been intimidated into silence? One scared officer reported anonymously to STARRS that he was threatened with "firing" if he even talked to STARRS. Why is the military denying his First Amendment rights?

What Is Indoctrination Doing to Our Armed Services?
Military members are voting with their feet. Three West Point cadets left the academy because of the COVID-19 vaccine mandate, increasing and troubling wokeness, and promotion of leftist ideas. The Army and Navy are now offering bonuses of $50,000 or more for joining in select fields, which is a sign of low recruitment—which the services have admitted is happening. Is this the Great Reset of the U.S. military? China and Russia are openly laughing at us and accelerating their imperialistic plans.

STARRS was formed to protect the U.S. military and our Republic from racism and radicalism. We never thought that the phrase in our oath about protecting against domestic enemies would reference this "woke" threat to our country and our way of life. We shouldn't reimagine our combat warriors into social-justice warriors. That is the definition of insanity.

This article originally appeared at AmericanThinker.com[55]

★ ★ ★

How Social Justice Is Killing the Military, American Thinker, 5 March 2022

How Implementing Diversity, Equity, and Inclusion Will Harm Readiness in the Armed Forces and Fail to Solve Anything

By Brent Ramsey

What are the Department of Defense's policies related to Diversity, Equity, and Inclusion, and Critical Race Theory? The Department of Defense is doggedly implementing the policies set in motion by an Executive Order (EO) issued by President Obama on August 18, 2011. When he signed Executive Order 13583—Establishing a Coordinated Government-wide Initiative to Promote Diversity and Inclusion in the Federal Workforce—the path was set to implement the Diversity agenda into our government, including the military. This EO was in force for much of the time Obama was in office and continued in force throughout President Trump's term. It was strongly reinforced and enhanced by EO 13985, issued by President Biden upon his inauguration. This new EO calls for Advancing Racial Equity and Support for Underserved Communities Throughout the Federal Government to strengthen the federal workforce by promoting diversity, equity, inclusion, and accessibility (DEIA). This EO builds and expands upon the EO that the Obama administration issued 10 years before. This EO says in part that its purpose is "To further advance equity within the federal government, this order establishes that it is the policy of my Administration to cultivate a workforce that draws from the full diversity of the Nation." Ignored is the fact that the United States is already the most diverse nation on Earth and that equal opportunity has been the law of the land for decades.

Thus, Diversity, Equity, and Inclusion (DEI) is the policy of the Department of Defense and has been for a long time. Exactly where the policy morphed from diversity and inclusion in Obama's 2011 order to the present-day diversity, equity, and inclusion is impossible to say for sure. Without question, the roots of these philosophies trace back decades in academia and even longer in the writings of educators, philosophers, and political activists, going back even to the French revolution and before. The teaching of Critical Race Theory (CRT) is not specifically mandated by these multiple EOs per se but is implicitly endorsed because DEI is the language of CRT.[56] According to Christopher Rufo, Senior Fellow at the Manhattan Institute:

"There are a series of euphemisms deployed by its supporters to describe critical race theory, including 'equity,' 'social justice,' 'diversity and inclusion,' and 'culturally responsive teaching.' Critical race theorists, masters of language construction, realize that 'neo-Marxism' would be a hard sell. *Equity*, on the other hand, sounds non-threatening and is easily confused with the American principle of *equality*. But the distinction is vast and important. Indeed, equality—the principle proclaimed in the Declaration of Independence, defended in the Civil War, and codified into law with the 14th and 15th Amendments, the Civil Rights Act of 1964, and the Voting Rights Act of 1965—is explicitly rejected by critical race theorists. To them, equality represents 'mere non-discrimination' and provides 'camouflage' for White supremacy, patriarchy, and oppression. In contrast to equality, "equity" as defined and promoted by critical race theorists is little more than reformulated Marxism."

There you have it! The promotion of equity has been going on for years and is now deeply embedded in DOD up and down the chain of command; it is institutionalized in policy, instruction, and practice, and it is the institutionalizing of Marxist ideology.

Why Is DOD Promoting CRT and DEI?

Obviously, the senior officials of DOD under both Presidents Obama and Biden are all political appointees. In both these administrations, these officials are true believers in the politics and priorities of Democrat platforms. So, they believe in DEI and CRT since these are now core beliefs of the left. Naturally, they will vigorously implement those policies as much as they can. Hiring and promotions within DOD for years have been slanted in favor of those who subscribe to DEI, and those who do not support DEI are marginalized, harassed, passed over for promotion, and otherwise canceled. It's no wonder that DEI has taken root, as it has had years to grow and is supported from within DOD by true believers.

During the first 5 years of the Obama/Biden years, an unprecedented number of flag and Commanding officers, almost 200, were relieved of command and sent packing.[57] Senior officers who were viewed as not in step with President Obama's policy were identified and purged.[58] It is surmised these removals eliminated a lot of institutional natural resistance to the implementation of DEI and CRT in the military. With many of the more senior officers now gone, more junior flags had to be promoted to higher rank sooner to fill the ranks. These officers were apparently more amenable, more malleable in accepting the ideology of DEI and CRT. Perhaps others lacked the moral courage to object, so they went along to get along and continue to get advancement. This might well be a deliberate strategy to eliminate potential opponents of DEI and CRT, install a higher percent of more liberal officers into the officer corps, combined with a lack of courage by some to stand up for the truth about these divisive ideologies and/or self-promotion; it resulted in these destructive policies taking firm root throughout the military.

It became evident that these influences had taken firm hold of the military, especially the academies, as proponents of CRT began appearing on academy campuses as early as 2017 when Ta-Nehisi Coates was invited to speak at West Point.[59] And, once most senior leaders had bought into the new views on DEI and CRT or were silent about what was going on, the normal top-down military chain of command saw to it that the ideas and policies were implemented throughout DOD. This was especially true at the service academies and other accession sources, as will be documented in greater detail later in this article. So entrenched were these ideas in the federal government, including the military, in 2020, President Trump directed banning the teaching of CRT in the government.[60]

What proof is there that the emphasis on Diversity, Equity, and Inclusion is actually happening?
Stand Together Against Racism and Radicalism in the Services, STARRS, Inc.[61] was formed earlier in 2021 for precisely the purpose of educating

the military community and America as a whole on what is going on in the military that has diverted it from its core mission of readiness into social-justice advocacy. STARRS has gathered incontrovertible evidence that this divisive threat is real. A sampling is provided below:

Warning Signs: Annapolis and the Navy

Powers That Be:

- DEI was directed to be implemented throughout the Navy.[62]
- Task Force One Navy promotes diversity, equity, and inclusion. Equity is the opposite of merit.[63]
- Chief of Naval Personnel advocates for bringing back selection-board photographs, a non-merit factor, to increase promotion selection for minorities.[64]

Those Hurt:

- Midshipmen lectured by radical Ibram X. Kendi. Kendi, the author of *How to Be an Anti-Racist*, advocates racism to make up for past racism.[65]
- CNO reading list requires all Navy personnel to read Kendi's racist book and also *Sexual Minorities and Politics*, a history of gay marriage becoming law, and *The New Jim Crow*,[66] a false promotion of the allegation that the U.S. criminal-justice system is inherently racist.[67]

Warning Signs: USAFA, Air Force, and Space Force

Powers That Be:

- Board of Visitors access to faculty, cadets, and information shut down by Department of Defense.

- Academy football coaches promote Black Lives Matter and, unwittingly, its radical, Neo-Marxian agenda via a video (later removed after nine months but still on YouTube).
- USAFA not responsive to[68] FOIA requests.[69]

Those Hurt:

- Cadets taught racist Critical Race Theory.[70]
- Academy Diversity Reading Room appears to be a one-sided view of American racial history.[71] Despite three requests and after more than six months, the Academy has not responded to a STARRS request for the list of the books in the room.
- Cadets force-fed mandatory indoctrination training on systemic racism and White privilege.[72]
- Top-rated LtCol. Lohmeier removed from command and separated from the Space Force without benefits because he wrote and spoke about Public Affairs-approved book about racism/radicalism spreading in Space Force.[73]

Warning Signs: West Point and U.S. Army

Powers That Be:

- Army 4 Star and Chairman of Joint Chiefs of Staff defends the study of Critical Race Theory and testifies to Congress he wants to understand "White Rage."[74]
- Army teaching seminars on CRT and White Rage.[75]
- Judicial Watch sues West Point for failing to comply with FOIA requests regarding teaching of CRT to cadets.[76]

Those Hurt:

- In 2020, the worst cheating scandal in 45 years at West Point involved 73 cadets, mostly athletes.⁷⁷

- Cheating was downplayed by West Point leaders, who used "equity" as an explanation of why the guilty were not expelled. Cheaters played in the Liberty Bowl in a disgraceful sports-matters-more-than-honor display.⁷⁸

Ta-Nehisi Coates spent two days at West Point in 2018 speaking about race. Coates's toxic views say all Whites are racist and calls for trillions in reparations for all Blacks.⁷⁹

In addition to the above examples, STARRS has compiled an extensive list of incidents involving cadets at the USAFA and their parents reporting on the divisive atmosphere now present at the academy. Examples include a separate diversity chain of command outside of the regular squadron chain of command, an atmosphere that is toxic to White Christians at the Academy, and a "better be careful about what you say" environment that keeps many cadets in fear and misery most of the time. Between complaints by cadets and parents of cadets and whistleblower complaints compiled by Senator Cotton's office, there are more than 600 legitimate complaints about divisive Diversity, Equity, and Inclusion indoctrination that is going on in the services.

How is Diversity, Equity, and Inclusion indoctrination hurting the military?

GAO has documented that readiness in the services is lacking.⁸⁰ The report found, "Every warfighting domain . . . is now contested as potential adversaries, most notably China and Russia, have developed and enhanced their own capabilities," according to the report. "The GAO found that reported domain readiness did not meet readiness-recovery goals identified by the military services." A report commissioned by Senator Cotton, Congressman Crenshaw, Congressman Gallagher, and Congressman Banks specifically singled out the emphasis on Diversity as a contributing factor in the Surface

Navy's poor performance and readiness issues over the past several years.[81] For example, the following statements are found in the report: "I guarantee you every unit in the Navy is up to speed on their diversity training. I'm sorry that I can't say the same of their ship-handling training."[82] And, "Sometimes I think we care more about whether we have enough diversity officers than if we'll survive a fight with the Chinese navy,"[83] lamented one lieutenant currently on active duty. "It's criminal. They think my only value is as a Black woman. But you cut our ship open with a missile and we'll all bleed the same color."[84]

The Heritage Foundation compiles one of the most comprehensive and respected reports in the nation on the state of our military. Its 2021 Report is alarming. Despite record spending of more than $700B, all the services are rated as "weak" or "marginal" with the exception of the Army as regards their ability to defend the nation against multiple threats. The signs are everywhere that the mission focus on warfighting is lacking. A Navy capital ship with a replacement cost of more than $3B burned pier-side in San Diego due to criminal conduct on the part of a Sailor (arson), negligence, poor leadership, and incompetence. Forty-six officials in the chain of command have been found culpable and will face punishment in one of the Navy's worst peacetime disasters in history. The damage was so extensive that the ship had to be stricken from the list of commissioned ships and sold for a pittance as scrap. Even more recently, the *USS Connecticut* collided at full speed with an undersea mount and suffered such severe damage that it was taken out of service and will need to undergo major repairs. The CO, XO, and Chief of the Boat were immediately relieved of their positions, due to the Navy's lack of confidence in their actions, which led to this disaster.

The Navy is desperately short of ships to face the astonishing and rapid rise of the PRC PLAN (People's Liberation Army Navy), where, in the USINDOPACOM AOR, our Navy is outnumbered 355 ships to 55—no, make that 54 ships. With all these readiness problems and leadership failures visible everywhere, how in good conscience can the leadership of the military pour millions of dollars and hundreds of thousands of hours into the pursuit of Diversity, Equity, and Inclusion in a force that is already the most diverse organization on the face of the planet? The Navy's Task Force One Navy documents that the Navy already has 43% minority representation versus a national demographic of 39%, and yet it is full speed ahead on even more diversity . . . for what purpose? For the purpose of better readiness? No, to satisfy a political agenda!

If DEI indoctrination succeeds, what impacts can we expect in the future? Military members are already voting with their feet. Recently three West Point cadets left the academy due to a combination of the Covid-19 vaccine mandate and its increasing and troubling wokeness and promotion of leftist ideas. More attrition at West Point is expected. On 14 November 2021, during an interview of the three former cadets about why they left West Point, retired West Point graduate COL Pat Hueman made notes of their comments about the atmosphere at West Point and shared them with this author:

- Political activism out of control
- Disillusioned cadets
- Angry parents
- Officials quibbling and lying
- Promotion of divisiveness
- Promotion of guilt
- Promotion of victimhood
- Refusal to accept accountability
- Obfuscation
- Ignoring FOIA and border laws. Numerous FOIA requests have been filed, with no answers.[85]
- "Political Commissars" installed to ensure leaders promote DEI[86]
- Graduates ill-prepared to defend our nation
- Lack of due process when dealing with vaccine-hesitant cadets who were denied counsel and harassed, berated, thrown off athletic teams, and subjected to abuse for asking questions about the vaccine.

- Active promotion of CRT by professors in multiple English, History, and Humanities classes.

- Cadets not allowed the protections of the Nuremberg Code.[87]

Retention is lagging as service members who do not wish to be subjected to racist indoctrination or White-supremacy nonsense are leaving the services in droves. A Marine LtCol recently told the STARRS President that "everyone is leaving." A recent graduate of Navy boot camp told the STARRS President that, even in boot camp, there was a lot of promotion of woke nonsense. Morale is down due to the unrelenting nature of the diversity, equity, and inclusion drumbeat, but there are also a lot of other unnecessary distractions, such as accommodating and having to pay for so-called "transgenders" to not only be allowed to serve but for the military to pay for their medical costs, including surgeries. Recruiting is likewise challenged—it was already challenged enough due to the increasingly shrinking pool of eligible young people who are even qualified to serve. By singling out White, Christian, heterosexual males for special scrutiny and indoctrination, the newly woke military is dooming itself to failure, as disproportionate numbers of new recruits come from these demographic categories. As proof that these things are going on, STARRS reports the fact that their repeated efforts to gain information on what is going on through the FOIA process are being illegally ignored by the Air Force.[88] The same thing is happening with the Army and Navy likewise refusing to release information on their educational programs or answer FOIA requests in a timely manner.[89] If the services have nothing to hide and are not indoctrinating our young officer candidates on things such as CRT and relentless DEI programming at the three military academies, why are requests for information not promptly answered?

The real truth about Diversity, Equity, and Inclusion in the military and in our society—other factors to consider.
Why do the leaders, the critical thinkers, the analysts, and the scholars of the Department of Defense accept the lies and distortions about race relations in America? Can these people be ignorant of some basic truths about our nation? They have easy access to the volumes of research over decades by the nation's top scholars, many of them Black, that explain the

cultural and political reasons for a small segment of the Black community to lag behind. Or do they just not want to know because the truth would put them in an awkward or uncomfortable situation in relation to the current leadership of the nation? So, what is the real truth about Diversity in the nation and in the military? Is DEI necessary to overcome racism—or what is now commonly called "systemic racism"—in America?

The truth is that American society, including the military, is overwhelmingly a meritocracy. Those who get ahead do so by virtue of their skills, qualifications, and hard work. Are we perfect? Of course not. But it is recognized around the world that the U.S. is the land of opportunity, freedom, and equal protection under the law. That is why millions strive to come here to join us. According to a 2021 survey conducted by CATO Institute, 91% of Americans are in favor of allowing for legal immigration. A racist nation would not feel that way, as immigration is overwhelmingly made up of ethnic minorities. The top five nations for legal immigration are Mexico, India, China, the Philippines, and Cuba. The top five nations for refugee admittance are Republic of the Congo, Iraq, Afghanistan, Myanmar, and Ukraine. In 2021, more than 23 million in America were foreign-born, mostly ethnic minorities. These are not signs of a racist society. Does "systemic racism," in fact, exist in America?

Is there still some residual racism in this country? Of course, but it is rapidly dying out all across America, despite the best efforts of race hustlers and politicians who would use race to divide us and whose constant cries of "racism" and "White supremacy" are an attempt to hold onto power and manipulate the American public into fear and division. The truth is not hard to find. Just look at the data. Interracial marriage is at an all-time high and increasing year by year. According to a 2021 Gallup poll, 94% of the public has a favorable attitude toward interracial marriage. This is not a sign of a racist country. Most minority families are in the middle class or higher. Those still stuck in poverty or lacking in education and opportunity are held back by cultural factors having nothing to do with racism. There are volumes of research and books that reject the flawed theories of systemic racism and CRT. Much of this scholarship is by Black scholars such as Dr. Thomas Sowell,[90] Dr. Shelby Steele,[91] Dr. Glenn Loury,[92] Dr. Wilfred Reilly,[93] Dr. John McWhorter,[94] Robert Woodson, Sr.,[95] and Dr. Carol Swain.[96] Is it possible that DOD leaders are ignorant of this tremendous body of work that explains why a small segment of the Black community lags behind? According to research by the Brookings Institute, disparities in a small percentage (less than 20%) of the Black population are cultural, not racist. Race has nothing to do with the reasons some are stuck behind. If one adjusts for just three factors—marrying before having children, finishing high school, and holding down a full-time job at any level, including entry level—success follows, regardless of race or other circumstances.[97] This result has been replicated by many studies and scholars, including top Black economists such as Thomas Sowell of Stanford and Roland Fryer of Harvard. Racism is not holding Americans back! Poor study, work, personal habits, and single-parent households, which the government exacerbates through pay-per-child welfare, are much more at fault for poverty and lack of opportunity than any residual racism that may exist in our nation. This is the message of 1776 Unites and scores of other Black-led public interest groups that, instead of making excuses for lack of achievement in the inner cities, urge young minority members to take responsibility for their own lives and success and not blame nameless and faceless racism on a nation that provides them every opportunity for success that any person should ever want. On top of that, Affirmative Action and college-admission policies have institutionalized giving a leg up to minorities for decades. Those who do not take advantage have only themselves and cultural established roadblocks to blame. Lack of success is not because of racism or discrimination in school admissions or housing, as those things are against the law and have been for decades. Finally, Dr. Glenn Loury, the Merton P. Stoltz professor of economics at Brown University, senior fellow at the Manhattan Institute, and visiting Fellow at Hoover Institution, explains bluntly what is going on with the alleged systemic racism that DEI is designed to overcome:

"The invocation of 'systemic racism' in political arguments is both a bluff and a bludgeon.

"When a person says, for example, 'Over-representation of Black Americans in prison in the United States is due to systemic racism,' he is daring the listener to say: 'No. It's really because there are so many Blacks who are breaking the laws.' And who would risk responding that way these days? The phrase effectively bullies the listener into silence.

"Users of the phrase seldom offer any evidence beyond citing a fact about racial disparity while asserting shadowy structural causes that are never fully specified. We are all simply supposed to know how 'systemic racism,' abetted by 'White privilege' and furthered by 'White supremacy,' conspire to leave Blacks lagging behind.

"American history is rather more subtle and more interesting. Such disparities have multiple, interacting causes, ranging from culture to politics to economics and, yes, to nefarious doings of institutions and individuals who may well have been racist. But acknowledging this complexity is too much nuance for those alleging 'systemic racism.'

"They ignore the following truth: that America has basically achieved equal opportunity in terms of race. We have chased away the Jim Crow bugaboo, not just with laws but also by widespread social customs, practices, and norms. When Democrats call a Georgia voter-integrity law a resurgence of Jim Crow, it is nothing more than a lie. Everybody knows there is no real Jim Crow to be found anywhere in America.

"The phrase also does a grave disservice to Blacks and to the country. Here we are now, well into the 21st century. (Have you heard of China?) Our lives are being remade every decade by technology, globalization, communication, and innovation, and yet all we seem to hear about is race.

"My deep suspicion is that these charges of 'systemic racism' have proliferated and grown so hysterical because Black people—with full citizenship and equal opportunity in the most dynamic country on Earth—are failing to measure up. Violent crime is one dimension of this. The disorder and chaos in our family lives is another. Denouncing 'systemic racism,' invoking 'White supremacy,' and shouting 'Black Lives Matter,' while 8,000 Black homicides a year go unmentioned—these are maneuvers of avoidance and blame-shifting.

"The irony is that so many of us decry 'systemic racism,' even as we simultaneously demand that this very same 'system' deliver us."

The statements of the scholars cited above are proof positive that the actions DOD is taking to promote DEI and CRT are misplaced, will not solve the culture problems or correct the public-policy mistakes of the federal government, and will have no favorable impact on race relations in the services. The opposite will be the effect of forcing DEI on the services, as many will see the flaws and will choose to go elsewhere rather than serve in such a divisive environment. That is already happening.

Conclusion: The keys to readiness, combat effectiveness, and victory over our potential enemies are a unified, motivated, skilled military team. That once-mighty team is being dismantled, piece by piece, day by day, inch by inch. Where unity and cohesiveness were once the norm, divisiveness, suspicion, lack of trust, enmity, and distraction have taken root. Napoleon famously said, *"In war, three-quarters turns on personal character and relations; the balance of manpower and materials counts only for the remaining quarter."*[98] Moral is defined as the human element, the human moral and ethical impetus to fight and win. The military is more divided, more fractured than at any other time in our history. The adoption of Diversity, Equity, and Inclusion is at fault for separating our fighting units into factions, groups, different identity units—even *opponents* and even when the nature of war cries out for unity and cohesion. If DEI wins, our nation is in grave danger from our enemies. And don't think our enemies are not aware of what we are doing to ourselves. Citizens everywhere must rise up and oppose this blight upon our military and expose the fallacies that lie behind them. The facts are at hand. What will you do to save our military? For more information on racism and radicalism in the military and how to fight to eliminate them, please visit www. STARRS.us. *This article appeared in CD Media.*

★ ★ ★

How Implementing Diversity, Equity, and Inclusion Will Harm Readiness in the Armed Forces and Fail to Solve Anything, Creative Destruction Media, 4 January 2022

Divide and Conquer: Radicalizing Military Education

Grievance-based curricula are coming to a military academy near you.

By John Cauthen

From late 2017 to 2018, three scholars wrote and submitted to academic journals 20 fake papers focused on gender and sexual identity—what the authors call "grievance studies." Seven journals published their submissions.[99] The scholars' stated objective for this experiment was to expose the farcical nature of these "disciplines" and their replacement of scholarly rigor with an overt political agenda devoid of academic seriousness.[100]

The hoax not only damaged the reputations of the publishing journals, but it also challenged the academic integrity of identity-based "grievance studies."[101] Such fields may well be part of universities' overall mission to expand areas of inquiry, study, and research. Reality, however, suggests that they are more about increasing staff and faculty diversity hires and enhancing political power than about perpetuating knowledge.[102]

Most external observers who are numbed by the absurdities of contemporary academia likely believe that the political damage emanating from identity studies is confined to the civilian Ivory Tower. In the five years since the hoax, however, the grievance chorus has breached the walls and escaped traditional academia. Corporations,[103] government,[104] and even the military[105] and military academies[106] are now true believers, a fact embodied by ubiquitous Diversity, Equity, and Inclusion (DEI) programs and offices.

If "grievance studies" lack academic rigor, and if some of their practitioners have been unmasked as charlatans rather than serious scholars, why have the nation's military academies so enthusiastically embraced their arguments and conclusions? The overall politicization of the military, an historically conservative and apolitical institution, is one potential explanation. As a recent Heritage Foundation report on military readiness asserts that civilian and military leaders are advancing "divisive progressive social justice ideologies" across the individual services, largely by viewing "all matters through the lens of DEI."[107]

Purveyors of identity studies are, of course, adherents and advocates of DEI, and the success of their progressive political agenda in America's other elite institutions—academia, media, government, and business—has renewed the objective of remaking the military in line with fashionable ideological views. Every federal military service academy, for example, has either established DEI offices or published strategic plans to expand this agenda. Increasingly, identity studies are required for some students, depending on their selected undergraduate major and minor.

For instance, a 2017 West Point memo requesting the creation of a diversity and inclusion minor (five courses and 15 credit hours) argued that the program of study would "leverage diversity and foster inclusion to prepare leaders of character . . . to effectively lead in a multicultural Army."[108] West Point[109] established its diversity and inclusion minor in 2020, followed by the Air Force Academy in 2021. The Air Force Academy had a similar objective, asserting that the program would develop "leaders who not only understand and recognize the importance of diversity, but who actively create inclusive environments that leverage this diversity toward mission success."[110]

On its face, none of this rather vague language is particularly problematic. A deeper examination, however, raises legitimate questions about how the coursework in question supports the stated mission of the Service Academies and furthers professional military education.[111] For example, some electives in these programs focus exclusively on race, ethnicity, gender, and sexuality—not exactly martial concerns.

A course offered at West Point as part of the diversity and inclusion studies minor is titled "The Politics of Race, Gender, and Sexuality" and places an emphasis "on the inherent inequalities found within the structures, rules, and processes of the American political system."[112] At the Air Force Academy, a course called "Gender, Sexuality, and Society" seeks to show that one's set of social "beliefs [can] create and enforce a system of difference and inequality."[113]

Although a modest sample, these examples are likely the first salvo in the campaign to increase identity studies across the military-educational ecosystem. While the Naval Academy does not currently have a diversity and inclusion minor, the English Department offers a 300-level course[114] titled "Gender and Sexuality Studies," with readings from Kimberlé Crenshaw, who coined the terms "intersectionality"[115] and "critical race theory."[116] Given the DEI push across the military, it is almost certain that this identity "scholarship" will grow at the Naval Academy to include a focused diversity and inclusion minor.

In fact, the Naval Academy and the smaller Merchant Marine Academy have already crafted the institutional plans necessary to accomplish these ends. The Naval Academy's Diversity and Inclusion Strategic Plan contains the stated goal of "partner[ing] with Academic Departments in conducting a comprehensive curriculum review prioritizing the inclusion of marginalized scholarship and hidden histories within midshipmen education."[117] The Merchant Marine Academy has a strategic plan that sets out to further entrench DEI via a comprehensive review of offered courses that will integrate "elements of diversity, equality [sic], and inclusion into the Educational Program for Midshipmen."[118]

Moreover, the Naval Academy's 2030 Strategic Plan includes an ominous pledge to "identify and address traditions, policies, and practices at USNA that support systemic bias."[119] It seems the administration has already decided that the Naval Academy is systemically corrupt and that the only remedy for this scourge is expanding the DEI agenda. This includes the likely expansion of identity-centric course offerings and programs.

Identity-based scholars, researchers, and advocates are promulgating their agenda beyond the confines of military undergraduate education. The U.S. Naval War College, for example, recently hosted the ninth annual Women, Peace, and Security Symposium.[120] The theme: "Share . . . knowledge on warfighting and conflict resolution focusing on the gender perspective."

Sessions included "Tackling Gender in Kinetic Operations" and "Politics of Belonging: Men as Allies in the Meaningful Inclusion of Women in the Security Sector."

Rear Admiral Stephen B. Luce, the founder of the Naval War College, stated in the 1880s that the college was to be "a place of original research on all questions relating to war and to statesmanship connected with war, or the prevention of war."[121] Given the nation's current geopolitical threats, does a gendered approach to studying tactics, operational art, strategy, and war really yield the greatest returns? Clearly, gender studies and other identity-based disciplines are about the DEI agenda, not about continued professional military education[122] and the study[123] of how to deter and defeat our nation's enemies.

DEI and identity-centric courses in civilian academic institutions are designed[124] to divide and sort people into ever-expanding, aggrieved, and oppressed identity groups. We should not expect a different experience for military students and personnel.[125] Where the military should be focusing on *esprit de corps*, critical thinking, tactical and strategic acumen, and developing an apolitical professional officer corps, it is instead aggressively embracing DEI and identity-based grievance studies.[126]

The outcome of this push remains uncertain, but a 2020 policy paper crafted by nine West Point graduates and addressed to West Point leadership may serve as a useful primer of what a fractured and Balkanized military might look like. Highlights include a demand that senior White leaders acknowledge that their "white privilege sustains systems of racism," as well as a call for West Point to release a statement acknowledging the "existence of anti-black racism endemic to West Point."[127]

Amazingly, the paper's authors advocate for self-segregation and the abandonment of colorblindness. "There is a need for space . . . that is specifically dedicated to black Cadets," they write. "The Academy must take action to dismantle the processes that safeguard white supremacy, to stop reinforcing color-blindness, and to invest in a space that preserves black Cadets' identity."[128]

Statements like these from civilian academics and students may not be surprising. That they came from active Army officers who took an oath to protect and defend the Constitution is shocking. If DEI and identity-based coursework become embedded throughout the military educational ecosystem, we should expect similar, if not more radicalized, beliefs by military members. By continuing to teach and encourage disunity, discord, and discrimination in the military ranks, our Armed Services may have already lost the next conflict without even firing a shot.

Republished courtesy of the James G. Martin Center for Academic Renewal

★ ★ ★

Divide and Conquer: Radicalizing Military Education was published by The James G. Martin Center for Academic Renewal on 07 June 2023

The United States: Most Racist Nation Ever

By Brent Ramsey and Michael D. Pefley

Those who allege the U.S. is a racist nation often cite income disparities between Whites and other ethnicities. That's the narrative that propels the Critical Race Theory initiative that has countless numbers of Whites groveling and suffering from irredeemable guilt. White fragility and anti-racism pressures have brainwashed the Woke into endemic guilt over their white skin. Are minorities at the bottom of the income scale due to systemic racism? Hardly! 2020 Census Bureau data shows, in stark terms, at least eight non-White ethnic groups do better than the average American, including Whites.

Can the paucity of mixed-race marriages be an indicator of a racist country? How can it be that racist America has a 10% rate of marriages that are of mixed race? This percentage has been steadily rising over the years. In 2021, 15% of all marriages were mixed race. And, according to Gallup, an overwhelming majority of 94% of Americans approve of mixed marriages.

The demographics of America are more racially diverse than ever, and this metric increases year by year. If this were truly a racist nation, Americans would not show up in droves every time to help their fellow citizens, no matter their race, after a disaster. America has 146 million people who identify as ethnic minorities. The continued lie that the U.S. is a racist nation is a repeated narrative from progressive groups, media, academia, and entertainment, and it isn't true.

If the U.S. is racist, how could we welcome so many people from different ethnicities, races, and cultures? You sure can't tell it from U.S. immigration policy. In the data from the Census Bureau, as of 2016, the foreign-born population was in excess of 43 million, of which only 8 million are White (18.6%). Why would America welcome millions of non-Whites from all over the world to our shores if we were racist through and through?

In the area of advanced degrees, 63% of the Master's degrees in 2020 were granted to Whites, and 37% were granted to minorities. According to the 2020 Census, the white/minority demographic was 61.3% versus 38.7%. The numbers above show that the American higher-education system is clearly not racist.

According to Brookings, the liberal think tank, the diversity of the nation's middle class matches the overall diversity of the nation. There has been dramatic progress over the past 40 years to the point where middle-class diversity is a representative overlay matching the national demographic percentages. Opportunity exists for our citizens to rise to the middle class and higher without constraints or bias due to racism holding people back.

On February 26, 2021, the U.S. Equal Employment Opportunity Commission (EEOC) released the Fiscal Year 2020 Enforcement and Litigation Data. Their statistics demonstrate the extreme rarity of discrimination in the U.S., contrary to the constant beating of the drum in the press about rampant racism. EEOC-received charges of workplace discrimination in FY 2020 was down roughly 7% from 2019. In drastic contrast, the agency received five times the number of charges in 2010. Where residual racism is found, it is investigated and adjudicated in accordance with the laws.

What about racism in holding public office as a member of Congress? At the seating of the 117th Congress, there were 124 minority members, a 97% increase in minority members in the House of Representatives since the Congress of 2001–2003.

With respect to law enforcement, 12.1% of the police in the U.S. are Black. The Black population of the U.S. stands at 12.5%.

Regarding military composition, Task Force One Navy reported that, in 2021, the U.S. military had a minority representation at 43% versus a national demographic of 39%.

The mainstream media, social media, academia, and entertainment industry are lying to the American public about "pervasive racism." Instead of telling the truth about how far we have come, how much racism is a thing of the past, how much opportunity and equality abound in our land, they lie, obfuscate, and parrot talking points about racism that are fabrications and not supported by fact.

There is an evil and dishonest breed of people who make their living from spreading lies about racism in America. Many are in academia, and they make their living telling lies to sell books and give highly compensated speeches about how racist the nation that they live in is, while they rake in millions of dollars. There are also many "race hustlers" in the media, and they have the same *modus operandi* as the academics, who label everyone that they disagree with as racist. This while they spread exaggerated stories about racism in a constant deluge to foment unrest and chaos. There is ample, legitimate scholarship about college admissions, employment opportunity, cultural factors by race, and the negative impacts of the welfare state on the state of the races in America today. The popular press promotes victimology for even the most entitled and richest Black people. For a reality check about true opportunity in America, read the scholarship of Dr. Thomas Sowell, Dr. Glenn Loury, Dr. Jason Hill, Dr. Shelby Steele, Dr. Wilfred Reilly, Dr. Carol Swain, or Robert Woodson, Sr.

There is a common myth promoted by the left and the media (Okay—that was redundant) that unarmed blacks are being preyed on by the police in large numbers. The Marxist Black Lives Matter (that has at the national-organization level done little to nothing for poor Blacks) movement is one of the purveyors of this lie. Actual DOJ data is confirmed by multiple independent sources, including the *Washington Post,* that the numbers of unarmed Blacks killed by police each year is tiny. An unarmed Black person is much more at risk of being killed by lightning each year than they are of being killed by a police officer. In fact, the data shows that a Black person is 400 times more likely to be killed by another Black than to be killed by a policeman. Another truth largely unknown and completely untold by the left and the media is that crime is largely either White on White or Black on Black.

In conclusion, too many Americans are deceived into thinking race relations and racism are big problems in the nation. As in George Orwell's *1984,* if a lie is repeated often enough, it becomes a *truth.* Those baseless stories are what the left, the media, academia, big business, and Hollywood want you to believe. Actual racism in our country is rare and getting rarer.

No one is allowed under the law to act on racist leanings or to harm/discriminate against another person. Discrimination through voter suppression has become a thing of the past. Those things are dead and gone, notwithstanding the left's and the media's attempt to bring them back to life. CRT, BLM, and DEI are all initiatives attempting to separate us. Americans are all equal under the law.

How long will it take for America to believe and shut down those who claim the USA is a racist nation?

This article originally appeared at AmericanThinker.com[129]

★ ★ ★

DEI Means the Death of Military Professionalism

By John Cauthen

Recent studies[130] and reporting[131] have highlighted the military's recruiting challenges for fiscal year 2022 with dire projections of future shortfalls in coming years. The reasons for recruiting deficits are many and complex, but one major factor is regressing civilian trust and confidence in the military. Since 2018, from record highs at the end of the last century and into the post-9/11 era, positive civilian impressions of the military plummeted.[132] This decreased trust is, in no small measure, due to bipartisan public perceptions of a politicized military from both the left and right.[133]

A 2022 Ronald Reagan Institute survey noted that half of respondents attributed military politicization to a "woke" agenda.[134] It is a sentiment certainly shared by some in civilian leadership and many current military members and veterans.

The Reagan Institute polling does not define "woke." Ipsos polling, however, suggests that the meaning of "woke" can be fluid.[135] Although "woke" may have different definitions, for many, it is associated prominently with Diversity, Equity, and Inclusion (DEI).

These DEI initiatives have penetrated most major American institutions[136] and are now openly and aggressively being implemented across the military services, from the highest echelons down to the unit level. The introduction of DEI into the military is a significant factor in the public's perception of an increasingly politicized Armed Forces.[137] Interestingly,

DEI is now being unmasked as inherently intolerant, despite nostrums of diversity and inclusion, most notably when someone's speech and ideas conflict with what those in the DEI apparatus deem appropriate thinking and behavior.

DEI's proliferation[138] in and increasing hostility[139] to academia[140] serve as a useful proxy to understand the public's concerns about a "woke" military. Stanford faculty members recently voiced concern about the stifling of speech and academic freedom via an anonymous bias-reporting system.[141] And a student mob at Stanford University's law school, enabled by a DEI administrator, recently shouted down an invited guest speaker, a federal judge,[142] hurling invectives inappropriate for any setting, let alone an elite law school. Because of DEI's threat, Harvard faculty felt compelled to create a Council for Academic Freedom "devoted to free inquiry, intellectual diversity, and civil discourse."[143] As Harvard Professors Steven Pinker and Bertha Madras explained, there is "an exploding bureaucracy for policing harassment and discrimination [that] has professional interests that are not necessarily aligned with the production and transmission of knowledge."[144]

It is no surprise that the public is skeptical of DEI in general and specifically its appropriateness for the military. DEI in the military will likely produce similar outcomes. The most effective way to counter this is to ensure that our military members, especially the officer corps, are being educated and trained to maintain and sustain professionalism consistent with historical civil-military norms.

The restoration of public trust requires the military to "examine how we educate servicemembers in the professional norms related to military service in a democracy."[145] At the Air Force Academy, thirteen cadets recently pressed for greater education in civics and civil-military relations to ensure they could "comprehend the professional norms required to support the Constitutional principles inherent in their military oaths."[146] This is crucial for developing and maintaining a professional and apolitical officer corps, a prerequisite for healthy civil-military relations. A focus on DEI detracts from this and corrupts professionalism through an overtly progressive political agenda.

There are few in and out of uniform who question the pre-eminence of civilian control, but this control does not require the military to become a facsimile of civil society or to import its programs, ideologies, and structures in part or wholesale. Samuel P. Huntington, author of *The Soldier and the State*, astutely observed in a later work that "the dilemma of military institutions in a liberal society can only be resolved satisfactorily by a military establishment that is *different from but not distant from* the society it serves."[147] This uniquely American civil-military dynamic will evolve, and each component will influence the other in search of equilibrium and institutional comity, but neither should disadvantage the other.

For the military to remain in balance with civilian institutions and provide security, it must maintain its uniqueness[148] while simultaneously responding to societal imperatives of civilian control.[149] Without this balance, said Huntington, "military institutions which reflect only social values may be incapable of performing effectively their military function."[150] DEI, primarily a civilian importation, is an example of precisely how a social value supported by few is inimical to military professionalism, unity, cohesion, and *esprit de corps*. The public recognizes this and is rightly aghast at the politicization of our military institutions by the introduction of DEI programs and offices across our military.[151]

Appropriate civilian control is also at odds with DEI. Again, Huntington's analysis of civil-military relations offers some useful insights, notably on the types of civilian control over the military and the natural tensions therein. "Subjective" civilian control, he argues, maximizes civilian power and is necessary when a professional military is absent. "Objective" civilian control "is that distribution of political power between military and civilian groups which is most conducive to the emergence of professional attitudes and behavior among the members of the officer corps."[152] DEI, because of its obsessive focus on taxonomic sorting by race and identity, is at odds with the idea of professionalism and contrary to the notion of objective civilian control so fundamental to our system of civil-military relations.

Perhaps one of Huntington's more prescient observations is this: "Subjective civilian control achieves its end by civilianizing the military, making them the mirror of the state. Objective civilian control achieves its end by militarizing the military, making them the tool of the state.... The antithesis of objective civilian control is military participation in politics:

civilian control decreases as the military becomes progressively involved in institutional, class, and constitutional politics."[153] In other words, optimal civilian control (objective control) exists when military professionalism is maximized. DEI hews more to the subjective vice objective archetype of civilian control. It is no surprise the public views DEI with wariness and military politicization as an aberration.

Strangely, recent commentators argue that this politicization is attributable only to the right. One former Naval officer, writer, and editor claimed that critics of "wokeness" and DEI are, in fact, agents bent on politicizing the military.[154] A political-science professor makes a similar argument, claiming that anti-woke resistance is "a rallying cry of the American right" and that those questioning its merits are actually endangering national security.[155] Apparently, progressives and their works are simply noble pursuits devoid of politics, and the right is the true threat.

Recent examples of DEI run amok in academia[156] and business[157] simply do not support these assertions. To contend otherwise is either naïve or willful blindness. DEI, conceived, birthed, and nurtured in a cloistered civilian academic environment, has no place in the military.

Taking note of the damage DEI has done to academia[158] and lack of value[159] provided to businesses, the public is right to be wary of DEI taking root in the military. DEI is political, and its programs, wherever they exist, seek to accrue and expand power through division rather than unity.[160]

Nothing about DEI is sympathetic to actual notions of diversity of inclusion. By embracing patently divisive DEI policies and programs, civilian and military leaders unnecessarily violate the public trust in the military's ability to remain an apolitical professional body devoted to national security. It is no surprise, then, why the public views DEI as an instrument to enact social change and press the prevailing political agenda of the day rather than fullfil the promise of diversity and inclusion.

Military professionalism and culture are unique. These attributes are also crucial for the health of civil-military relations. "To despise the military," said political scientist Herbert Garfinkel, "is to open the way for an oppressive form of civilian supremacy that would ultimately endanger our democratic values by leaving the nation effectively defenseless."[161] We would do well to ponder this quote as we debate the wisdom of policies and programs like DEI in the military, the education of its personnel, and the development of a professional officer corps so critical to our national security.

This article originally appeared at AmericanThinker.com[162]

★ ★ ★

DEI Means the Death of Military Professionalism was published by American Thinker on 21 April 2023

Task Force One Navy Final Report: "The Emperor's New Clothes" Redux

By Phillip Keuhlen

Executive Summary

The Task Force One Navy Final Report was launched from the twin assumptions that the Navy suffers from systemic racism and that diversity benefits the Navy's military mission. Those assumptions are not supported by evidence. The use of data in support of the report's findings and recommendations is flawed by logical fallacies and by misapplied or misrepresented data to support its distorted narrative. Themes that are central to Critical Race Theory and its allied movements find an active voice in the TF1N Charter and Final Report, confirming a philosophy and a political agenda on the part of those who wrote the Task Force One Navy Charter and steered the enterprise. The fundamental question posed by the issuance of this report is whether the Navy's leadership is placing the Navy's readiness to defend the Nation at risk (and unwittingly subverting the Nation's Constitution and laws) by jettisoning its colorblind, meritocratic culture and entering a significant transformation without a rigorous, fact-based, logical justification.

Introduction

In the wake of racially motivated civilian riots in the spring and early summer of 2020, the U.S. Navy's Chief of Naval Operations stood up Task Force One Navy (TF1N) on July 1, 2020.[163] After a six-month effort, the final 142-page report was submitted on January 28, 2021.[164] The initial charter proceeded from the key assumption that the Navy, as an institution, is systemically racist.[165] The TF1N Final Report Executive Summary identified a second key assumption guiding TF1N, stating, "Mission readiness is stronger when diverse strengths are used and differing perspectives are applied."[166] In the context of what TF1N analyzed and reported on, one can only understand "diverse strengths" and "differing perspectives" as meaning demographic diversity arising in race, ethnicity, sexual orientation, and biological gender.

Not surprisingly, considering the key entering assumptions, the Final Report identified systems, climate, and culture problems, and submitted almost 60 recommendations aligned with four Lines of Enquiry (LOE): Recruiting, Talent Management/Retention, Professional Development, and Innovation and STEM. A fifth LOE, "Additional," was added for additional recommendations not fitting neatly into the original LOEs.

However, one should be skeptical about the entire exercise of TF1N and the recommendations that flow from it. It inaccurately depicts the proud institution of the United States Navy as systemically racist, a slander that has more potential to undermine morale, good order, discipline, and military effectiveness than any recent national adversary's actions. **There are several reasons to consider it fundamentally flawed in its approach, inaccurate in its portrayal of the current status of the Navy, and blatantly misrepresentative of the achievement of equality of opportunity in the service through the efforts of generations of Naval personnel.** Those reasons include:

- Key assumptions are not supported by evidence.
- Some conclusions are driven by faulty assumptions.
- Some conclusions are the result of logical fallacies.
- Some data is misapplied or misrepresented.
- Its language clearly indicates a political agenda.

The Military Value of Diversity

In assessing the current state of the Navy, the TF1N Final Report states that the importance of diversity to the service rests in improved performance by diverse teams.[167]

"These statistics are important because diverse teams are 58 percent more likely than non-diverse teams to accurately assess a situation.[168] In addition, gender-diverse organizations are 15 percent more likely to outperform other organizations, and diverse organizations are 35 percent more likely to outperform their non-diverse counterparts."[169]

Promoted Content
These are impressive claims, but beyond these top-level summary statements, what do the cited source documents actually say in their details, and what are their implications for the U.S. Navy? These are important questions since these are the *only* performance-based evidence presented by the TF1N to support its entering assumption that "Mission readiness is stronger when diverse strengths are used and differing perspectives are applied."

Proceedings of the National Academy of Science **Article**
The Proceedings of the National Academy of Sciences of the United States of America (PNAS) is a peer-reviewed multidisciplinary scientific journal. The article cited by TF1N is concerned with the effect of ethnic diversity on a misfit between market prices and the true value of assets (a bubble) in sets of ethnically diverse and ethnically homogeneous traders.[170] The authors conducted the study because previous research on the effects of ethnic diversity in economic growth, social capital, cities and neighborhoods, jury deliberations, organizations, and work teams *were equivocal*. Some studies found benefits, while others did not, a critical point not presented or considered in the TF1N Final Report.[171]

The authors conducted stock-trading simulations in two markets, Southeast Asia (Chinese, Indian, and Malay participants) and North America (White, Latino, and African-American participants), with individuals having the requisite financial skills. All trading information was true, public, and anonymous, and all participants could see all completed transactions and bid/ask offers. The trading accuracy of each participant was evaluated beforehand and considered in evaluating results.[172]

The authors concluded that "Across markets and locations, market prices fit true values 58% better in diverse markets."[173] This summary is the only factoid from this entire study cited by TF1N. But the authors also analyzed the *cause* of the performance difference they observed and its implications.

"Ethnic diversity facilitates friction. This friction can increase conflict in some group settings, whether a work team, a community, or a region. Conversely, ethnic homogeneity may induce confidence, or instrumental trust, in others' decisions (confidence not necessarily in their benevolence or morality, but in the reasonableness of their decisions, as captured in such everyday statements as '*I trust his judgment*'). However, in modern markets, vigilant skepticism is beneficial; overreliance on others' decisions is risky."[174] (Emphasis added)

In this experiment, the gains come from increased friction and reduced mutual confidence, not from any cognitive diversity or assortment of perspectives and skills commonly posited to improve problem solving via enhanced diversity. This insight begs several questions.

- Is the accurate determination of the value of stocks in a commercial market strongly analogous to military operational decisions at the tactical unit level or strategic national level? TF1N does not address this question. This is of vital importance to the relevance of any insight from the study, given the authors' acknowledgment that the extant literature is equivocal at best on the benefits of diversity.

- What is the relative importance of any putative performance improvement from diversity in a military setting compared to facilitating friction, increasing organizational conflict, and reducing instrumental trust in others' decisions, which the authors assess as the cause of the improvement noted?

- On what basis did TF1N extend the narrow finding of the study ("market prices fit true values 58% better in diverse markets") to a much more expansive "diverse teams are 58 percent more likely than non-diverse teams to accurately assess a situation."

Considering the two key assumptions of TF1N identified previously, it seems reasonable to conclude that the TF1N authors cherry-picked a number from the cited study to support one or both of those assumptions.

They misrepresented the study's applicability as much broader than it is. They ignored its conclusions about causality that are, at a minimum, adverse to unit cohesion, if not good order, discipline, and military effectiveness.

McKinsey & Company Report
McKinsey & Company is a worldwide management-consulting firm that advises on strategic management to organizations across the private, public, and social sectors. According to their website, their focus is on organizational transformation and is informed by a deep commitment to diversity and making a positive social impact through their work.[175] The "Diversity Matters" report cited in the TF1N Final Report looked at the level of diversity (defined as a greater share of women and a more mixed ethnic/racial composition in the leadership of large companies) and company financial performance (measured as average EBIT[176] 2010–2013).[177] The Executive Summary of this report is the source of the TF1N Final Report assertion that "In addition, gender-diverse organizations are 15 percent more likely to outperform other organizations, and diverse organizations are 35 percent more likely to outperform their non-diverse counterparts."[178,179]

The McKinsey report was prepared using a proprietary database of information on 366 commercial companies in 7 industry groups. Top management teams were taken as defined in company websites and comprised more than 5,000 individuals, from the C-suite to senior vice presidents. Ethnic and racial categories were African Ancestry, European Ancestry, Near Eastern, East Asian, South Asian, Latino, Native American, and others.[180] It is important to understand that the McKinsey report was prepared from proprietary sources that are not shared and cannot be checked, unlike a peer-reviewed academic paper. In addition to these limitations, McKinsey & Company is explicitly committed to diversity and is engaged in selling its diversity-related expertise to other companies as transformative. These circumstances make this document more akin to a marketing brochure than serious scholarship. Nonetheless, the authors do point out some features that may affect the value of the report, while others are readily apparent. These include:

- The analysis applies directly only to the association between executive team and board of directors diversity and the measure of EBIT in selected commercial sectors.

- "The relationship between diversity and performance highlighted in the research is a correlation, not a causal link."[181]

- The observed relationship between diversity and EBIT results varied significantly by region (UK vs. US/Canada vs. Latin America), by "flavor" of diversity (Gender vs. Racial/Ethnicity), and by whether it occurred in the executive team or in the board.[182]
    – Overall, gender diversity was a much more potent correlate of increased EBIT margin than ethnic diversity.
    – In the U.S. and Canada, gender was *not* a statistically significant correlate of increased EBIT margin.

- The use of proprietary data precludes independent review and analysis of McKinsey's report, the hallmark of scholarly research.

As with the PNAS paper discussed previously, using the McKinsey report by TF1N invites critical observations and objections.

- How is the improvement in EBIT correlated with the diversity of only the most senior executives and directors in commercial markets strongly analogous to military staffing at all levels and operational decisions at the tactical-unit level or strategic-national level? TF1N does not address this question that is at the crux of the relevance of any insight from the study.

- Why did the TF1N Final Report misrepresent the finding of the McKinsey report that "gender-diverse organizations are 15 percent more likely to outperform other organizations" as applicable to the U.S. Navy, when the report explicitly shows that in the U.S. and Canada, gender was *not* a statistically significant correlate of increased EBIT margin?

- Why did TF1N ignore the clear observation in the McKinsey report that correlation does not confirm causation and present its "reasonable hypotheses" as fact?

- On what basis did TF1N extend the narrow applicability of McKinsey's study ("the level of diversity" [defined as "a greater share of women and a more mixed ethnic/racial composition in the leadership of large companies and company financial performance"]) to the much more expansive, "In addition, gender-diverse organizations are 15 percent more likely to outperform other organizations, and diverse organizations are 35 percent more likely to outperform their non-diverse counterparts"?

Besides its own research, the McKinsey paper summarizes outside sources supporting its assertions that diversity results in improved decision making and innovation. While that information is not cited directly in the TF1N Report, it is instructive to examine whether McKinsey uses the resources rigorously in its marketing environment. Since the TF1N Report focused on the decision-making benefits of diversity as its primary rationale, the section of the McKinsey report entitled "Better Decision Making and Innovation" was examined. This section is meant to bolster support for McKinsey's main assertion of gender and ethnic/racial diversity's value in making better business decisions. In that respect, it is similar to the TF1N's use of citations of external resources to bolster its claims for the value of diversity.

- McKinsey uses research by Page that is focused on professional (i.e., cognitive) diversity of "informed views" among executives (e.g., engineering vs. legal vs. financial backgrounds, rather than demographic diversity) to support its assertion of the value of gender and ethnic/racial diversity. The assertion is unsupported by hard evidence of any relationship between the two very different types of diversity.[183] In fact, the results Page documented are much better explained by scholarship in cognitive theory, the attributes of high-performance/high-reliability organizations, and research on methods to build and sustain them.

- McKinsey quotes anecdotally from an interview with a CEO in the food-sweetener business who notes, "Diversity creates dissent, and you need that. Without that, you're not going to get any deep inquiry or breakthroughs."[184] The level at which such dissent is beneficial is not discussed, and the cost/benefit balance of organizational dissent outside the context of innovation is ignored.

- McKinsey quotes from a *Harvard Business Review* article and an associated study from the Center for Talent Innovation to state that, "In short, the data strongly suggests that homogeneity stifles innovation." However, the nature of the study and the strength of the results are underwhelming. The study "data" is solely the result of working-level employee surveys on whether they felt that their ideas were accepted and acted upon by management. It correlated those results (again, correlation, rather than causation, is demonstrated) with management diversity. It then extrapolated causality to recent gains in market share or capture (as perceived/reported by the surveyed employees, rather than documented from public records). A modicum of managerial experience shows that the majority of employees know little about the formal or informal cost/benefit analyses that guide whether employee suggestions are advanced or abandoned and, often, less about the financial or market performance of their companies. The suggestions of causality in this study are very weak, at best.

- Finally, the McKinsey study cited the PNAS study discussed in more detail earlier. McKinsey generalized its results to, "If a market is dominated by any one ethnicity, it tends to make worse decisions."[185] This clearly misrepresents the study's very limited scope and applicability as a much more expansive interpretation.

TF1N Demographic Analysis

We see from the preceding sections that one of the key assumptions in the TF1N Final Report—of a performance advantage accruing to organizations solely on the basis of demographic diversity—rests on a tenuous foundation. Where it is based in research, the research is extrapolated far outside of its basis (PNAS & McKinsey) or rests upon research that does not meet common standards for scholarly inquiry (McKinsey). The TF1N Final Report completely ignores clear warnings in the sources it cites about equivocal findings of diversity in the literature, potential adverse effects due to demographic diversity (of a type that should be of concern for a military organization), and the logical fallacy of equating correlation with causation. The TF1N Final Report misrepresents and overextends its source materials to present a key assumption for its effort as fact. This, coupled with the Charter's other assumption that the U.S. Navy is systemically racist, strongly indicates that the TF1N team started with an answer in search of justification rather than as an honest exercise in self-assessment.

Beyond the fundamental consideration this paper raises about whether focusing on increasing gender and racial/ethnic diversity is a net military benefit to the Navy, the TF1N Final Report provides no credible rationale for the selection of the demographic makeup of the country as a whole as the yardstick against which it selectively gauges measurement of diversity in the service.[186] This is critically important because the chosen yardstick has the potential to radically shift the outcome of the assessment (if one assumes that the outcome is not dictated in advance). So, let us examine why this particular basis of measurement is not the proper yardstick and how that choice affects the analysis if more appropriate bases are used.

The Navy has no control over the processes that bring individuals into the potential candidate pool for military service, nor may the Navy recruit from the entire population pool. Using U.S. Census estimates,[187] we find that, in 2018 (the year used by TF1N for its data), only 36% of the U.S. population is of military service age, 18–44 years old. Further, with good reason, the Navy is enjoined from considering the entire pool of military-age individuals for accession to the service. For instance, there are minimum standards of physical fitness and educational attainment. Also, prior activities reflecting moral turpitude or potential vulnerability to hostile powers (such as criminal convictions or financial difficulties) may be disqualifying. In short, the demographic from which the Navy can recruit is not the demographic of the country as a whole. If, without agreeing that it has benefit to do so, one is to judge the diversity of the Navy against any basis of measurement, then it can only be intellectually honest to judge its performance against the demographic diversity of the smaller pool of qualified candidates available to it by law and regulation, rather than to the population as a whole.

But do these requirements mean that the demographic diversity of the qualified-candidate field could deviate substantially from that of the country as a whole? Fortunately, information is also available from U.S. Census sources to estimate the demographic profile of the Navy's qualified-candidate pool using the key requirement attributes of age and educational attainment.[188,189]

Examining TF1N's assessment of the current state of the U.S. Navy, we find the following data used to support the assessment of diversity problems in the officer corps in particular.[190]

Demographics (2018 Census)		Officer	Enlisted
White	76%	77%	59%
Black	13%	8%	19%
Asian	6%	6%	6%
Hispanic Ethnicity	18%	9%	18%
Female	51%	20%	20%

While the 2018 demographic percentage accurately represents the U.S. Census Bureau estimates for the entire U.S. population, that is not the demographic distribution of the population from which the Navy can recruit. Let's examine what happens when we adjust the basis of measurement to the population pool eligible for recruitment.

When one adjusts the demographic basis of measurement to consider age (18–44 years) and educational attainment (HS diploma for Enlisted, bachelor's degree for Officers) to evaluate the Navy's demographic diversity against that of that portion of the population that is eligible to serve, a radically different picture appears.[191]

Measured against the demographic diversity of the *eligible* pool of candidates, the Navy is succeeding, *not* failing. The officer corps is within +

3% of the eligible population in all categories (White, Black, Asian under, Hispanic, Female over), with no gender gap, while the enlisted community has notably overachieved in attracting Black and Hispanic Sailors relative to their percentages in the eligible pool. The *only* enlisted demographics that have significant underrepresentation relative to the eligible pool of candidates are Whites and women.[192]

Eligible Pool Demographics (2018 Census)		Enlisted	Eligible Pool Demographics (2018 Census)		Officer
White	70%	59%	White	79%	77%
Black	12%	19%	Black	9%	8%
Asian	6%	6%	Asian	9%	6%
Hispanic Ethnicity	11%	18%	Hispanic Ethnicity	7%	9%
Female	52%	20%	Female	19%	20%

This situation of *actual* relative success versus *portrayed* relative failure continues through the TF1N current assessment, particularly regarding LOE 1, Recruiting, and its discussion of underrepresented community accessions. As demonstrated in the tables below, when accession's demographics are properly measured against the demographics of the population pool *eligible* for accession, the fiction of underserved minority communities evaporates.

Eligible Demographic (2018 Census)		TF1N Metric Basis (2018 Census)		FY19 Officer Recruits	USNA 2019	USNA 2023	NROTC FY19
Racial Minorities: (Black + Asian)	18%	Racial Minorities: (Black + Asian)	19%	25%	23%	27%	17%
Hispanic Ethnicity	7%	Hispanic Ethnicity	18%	12%	12%	13%	10%
Female	19%	Female	51%	26%	28%	26%	19%

Eligible Demographic (2018 Census)		TF1N Metric Basis (2018 Census)		FY 19 Enlisted Recruits
Racial Minorities: (Black + Asian)	18%	Racial Minorities: (Black + Asian)	19%	25%
Hispanic Ethnicity	11%	Hispanic Ethnicity	18%	18%
Female	52%	Female	51%	24%

Eligible Demographic (2018 Census)		TF1N Metric Basis (2018 Census)		Officer Accessions
White	79%	White	76%	75%
Black	9%	Black	13%	7%
Asian	9%	Asian	6%	7%
Hispanic Ethnicity	7%	Hispanic Ethnicity	18%	11%
Female	19%	Female	51%	22%

For the Officer corps, virtually all commissioning sources comfortably exceeded the racial, ethnic, and gender distributions in the population that were eligible for accession. The sole near miss was a 2% shortfall in one year of NROTC racial-minority accessions. For Enlisted Sailors, racial and ethnic accession demographics again comfortably exceeded the relevant distributions in the population that is eligible to enlist. Only regarding the gender of Enlisted sailors is there any underrepresentation.

This is not the only place in which the TF1N Final Report suffers from inappropriate choices for its basis of measurement or from misrepresentation of the statistical significance of the result it presents. In its Navy Wide Advancement Exam data assessment, TF1N states that for E-4 to E-6, "minority advancement rates are lower compared to their white peers."[193] This conclusion is inaccurate in its portrayal because it ignores demographic data for the "other/multiple races" category. This category is typically about 8–10% of the pool considered for advancement. One can compute the "shortfall" of Minority selectees by comparing the actual number selected to the number that would have been selected at the average selection rate. When one does this, invariably (including in E-7, E-8, and E-9 selections), the total "shortfall" is of the same magnitude as the number in the "other/multiple races" category. More often, the Minority selectee "shortfall" is dwarfed by a pool of "other/multiple races" selectees that is 2–3 times greater than the assessed shortfall. When all demographics are considered, the narrative of lower minority-advancement rates evaporates.

As in the discussion of accessions,[194] TF1N notes—but does not identify as an issue or flag for corrective action—actual inequities in Minority-enlisted advancements that do not fit its narrative of systemic racism. For instance, neither the observation that "racial minorities accounted for higher percentages in the Meritorious Advancement Program (MAP),"[195] nor that "data for E-7 and E-9 during the most recent four-year period (FY17-FY20) shows that Black minorities and Hispanic ethnicity advanced higher than their white peers"[196] elicited any concern about those inequities or recommendations for addressing them.

The TF1N treatment of minority-officer promotions are similarly disingenuous. It continues its pre-selected narrative of overrepresentation of White officers in selection results. In so doing, it ignores two factors previously discussed and one new consideration, all of which contradict its narrative. These are:

- Using the entire U.S. population as a basis for comparison, rather than the portion of the U.S. population that is of military age and has a bachelor's degree. The previous discussion applies. The Navy data does not show White officers being selected at percentages different from their representation in the population eligible for commissioned service by age and education.

- Continuing to ignore the demographic category of "multiple race codes/other" creates a strongly distorted portrayal of a nonexistent "shortfall" in Minority-officer promotions. The previous discussion applies—that is, when that portion of the officer population that is coded as "multiple race codes/other" is considered, the underrepresentation of Minority officers in promotions evaporates.

- Ignoring the problem of "small numbers" in comparing selection percentages in the Flag ranks. The TF1N Final Report identifies the outsized effect that small changes in selection numbers can have[197] but completely ignores how that effect, coupled with the issues above, bears upon its conclusion that White officers are overrepresented in the flag ranks. In point of fact, over the five-year period considered by TF1N,[198] a shift of only one selectee per year between White and any other demographic category results in underrepresentation of White officers in the Flag Ranks, clearly an unwelcome insight for the authors of the TF1N Final Report.

The TF1N Agenda

When one sees a study begun with explicit assumptions that make its outcome preordained and its use/manipulation of data an exercise in bias confirmation, one is forced to ask why the effort was undertaken at all. In the case of the TF1N Final Report, the frame of reference and the agenda are evident in the language and terminology used throughout. The clear purpose of the TF1N report is to justify and muster support for the Navy's diversity initiatives, with the words "diverse" and "diversity" frequently appearing, on average, more than twice per page throughout the report. One might be excused for wondering why the leadership of an organization that is as self-evidently diverse (e.g., by gender, race, age, geography, socio-economic status, religion, education, national/ethnic origin, etc.), as the United States Navy is focused so singularly on diversity. The immediate answer to this lies in its precursor, Executive Order 13583 of August 18, 2011,[199] that established a "Coordinated Government-Wide Initiative to Promote Diversity and Inclusion in the Federal Workforce." One of the key features of this Executive Order was its philosophical framework. The Executive Order directly discusses the importance of equal opportunity and merit in the implementation of the order and recommits to those principles, stating the "order shall be implemented consistent with applicable law," explicitly citing 5 U.S.C. 2301. Indeed, within the framework established in Executive Order 13583, there are bona fide reasons bearing upon the Navy's core military mission to assure that personnel policies attract, retain, and promote adequate numbers of the best-qualified personnel. Some of the TF1N recommendations appear to be aimed at such goals. But the language used in both the TF1N Charter and the TF1N Final Report is replete with politically charged language that is clearly unnecessary if the sole purpose of the effort is optimizing personnel policies associated with military readiness.

A useful reference in this regard is Richard Delgado's seminal work *Critical Race Theory*,[200] which contains both a Glossary of Terms[201] and a discussion of the basic tenets and themes of Critical Race Theory (CRT).[202] It is essential reading for any citizen interested in understanding the currents and countercurrents of contemporary American civic discourse. Familiarity with this work shows that themes that are central to CRT and its allied movements[203] find an active voice in the TF1N Charter and Final Report. Central to the TF1N narrative is demonstrating the existence in the Navy of one of the key themes of CRT, the underrepresentation of minorities in the distribution of social goods, benefits, privilege, and status. The term "underrepresented" repeatedly occurs in the TF1N Final Report (mentioned more than 40 times). Indeed, the TF1N Final Report is extensively laden with the language and jargon of CRT, including "lived experience," "intersectional experience," "systemic inequities and racism," "structural and interpersonal biases," "equity," "unconscious bias," and "identity." *The incorporation of this language cannot be an accident: it does **not** appear in the source Executive Order. At a minimum, it confirms a philosophy and a political agenda for those who wrote the Task Force One Navy Charter and steered the enterprise.* And it provides the answer as to why data was manipulated in support of the inequities narrative: without the mathematical sleight of hand to create the false impression of underrepresentation, the entire effort collapses.

Critical Final Questions
Critical questions remain. First, does the Navy leadership understand the philosophy underpinning TF1N and the political intent inherent in the effort? Do they share that political agenda, or are they simply uninformed, unconcerned "useful idiots"?

The reason the answer to this question is critical is that, as Richard Delgado points out, *"critical race theory questions the very foundations of the liberal order, including equality theory, legal reasoning, Enlightenment rationalism, and neutral principles of constitutional law"* [204](Emphasis added). In short, CRT rejects the founding principles of the American nation reflected in the Declaration of Independence and defended by the Constitution that every Navy Sailor and Officer has sworn an oath to support and defend. CRT seeks to rewrite the history of the country,[205] replace its principles of governance, and reorder civic relations. CRT would replace colorblind equality of opportunity with equity—equal outcomes based on racial identity.[206] CRT would abandon the model of assimilation to American customs, measures, and laws that were considered vital to the survival of the Republic by Washington, Madison, and Hamilton, endorsing separatism and segregation.[207] CRT rejects the Declaration's assertion of rights[208] and the Constitution's ideal of equal protection under the law, and endorses aggressive color-conscious governance and jurisprudence.[209] CRT disputes the practices of merit and academic standards as inherently racist.[210] When the TF1N Charter and Final Report embrace the racialist philosophy of CRT, they depart from the guidance of Executive Order 13583. As recent statements by the Chief of Naval Personnel indicate,[211] they point at explicitly considering race in officer promotions contrary to the guidance of 5 U.S.C. 2301 that "selection and advancement should be determined *solely* (emphasis added) on the basis of relative ability, knowledge, and skills after fair and open competition which assures that all receive equal opportunity."[212]

The fundamental question posed by this analysis is whether today's Navy leadership is jettisoning the Navy's previously colorblind, meritocratic culture (and unwittingly subverting the nation's Constitution and laws) without a fact-based, logical justification capable of withstanding rigorous scrutiny? In the process and actions they have launched with the TF1N Final Report, do they place the Navy's readiness to defend the nation at risk?

★ ★ ★

Task Force One Navy Final Report, Real Clear Defense, 6 December 2021

Why This Matters

IN THIS CHAPTER, WE EXPLORE why this situation matters, not just to us but to the nation as well. Honor has always been a part of military service. Honor and trust go hand in hand. You cannot have an effective military unit without trust. The captain of a ship has to trust his officers, especially his deck officers and other key watch officers. Likewise, the wardroom and the crew have to trust the Captain that he or she is knowledgeable, competent, and a capable leader. Diversity, Equity, and Inclusion (DEI) policies introduce questions about how leaders were selected and the basis for promotion, thus undermining trust. Most DEI programs involve a separate reporting structure outside the chain of command, again undermining trust and confidence.

We have to know that our leaders are honest with us. This isn't a game of politics, where we take whatever politicians say with a grain of salt. Honor is based on truth. You have to be truthful with your subordinates, and you have to be truthful with your superiors. You might get by once or twice being untruthful or stretching it, but the troops will know, and your credibility and trust all goes to pot. The COVID pandemic and policy of forced vaccines and subsequent exposés of false government policy have contributed to the decline in trust for those serving and has harmed recruitment as well.

"Qualifications of a Naval Officer," once attributed to John Paul Jones but later learned to have been Written by Augustus C. Buell in 1900 to reflect his views of John Paul Jones, speaks of honor explicitly. Honor, trust, and capability are all things that matter. True, they matter in civilian life as well. But in a military situation, they matter more—lives are at stake. How much confidence do you have in the transgender person on watch who is undergoing treatments which may affect his/her/their ability to focus and carry out their duties? Are they more concerned about themselves or about their shipmates? "One hand for the ship, the other for the man" is a longtime catch phrase for mariners.

Many veterans continue their connection with their prior service organizations and are discouraged by what they see and hear. Stories abound wherein veterans are advising their children and those of friends and relatives not to seek a career in the military services. Military services were once considered the most trusted group of people in the nation. In 2022, reporting on trust, Pew Research noted, "Other prominent groups—including the military, police officers, and public school principals—have also seen their ratings decline." Our nation and our liberty depend on the military, and loss of trust is a serious matter to all.

★ ★ ★

Now Hear This

By Brent Ramsey

"Now Hear This" is the classic alert in the Navy that something important for all hands is about to be announced. This article is a "Now hear this!" for every American.

Our Navy is "*in extremis*," a nautical expression that indicates imminent danger. The *extremis* condition applies to the entire United States Navy and thus to the nation, as we are fundamentally a maritime nation.

Virtually everyone agrees that the biggest threat to the U.S. today and our way of life is from China. As far back as 2000, Congress, to its credit, recognized that China was going to be a threat and created the US-China Economic and Security Review Commission. The Commission produces an annual comprehensive report, typically running 600+ pages, on every aspect of the threat posed by China. In the 2023 report, as we came out of the Covid pandemic, we learned that things are becoming even more threatening. The latest report says, "The CCP gave its leader Xi Jinping unprecedented power over the Party and the country. Xi and the CCP relied ever more heavily on nationalist appeals, as was evident in its escalating rhetoric and menacing military actions toward Taiwan. Faced with a series of crises and unexpected developments, China's Communist Party regime reacted, not by reexamining its assumptions and modifying its approach, but rather by doubling down on existing policies. In the near term, these choices have increased the challenge China poses to the security, prosperity, and shared values of the United States and its democratic allies and partners."

A significant part of countering external threats from China falls to the Navy, as the most serious threats manifest themselves in maritime scenarios, especially in seas adjacent to China. Recognizing this threat from the PRC is why the largest combatant command is INDOPACOM. Fully 60% of our forward-deployed Naval combatants are in the Pacific, keeping watch on China.[213] Fact 1: The world is 70% ocean. Fact 2: 90% of the world's commerce goes by sea. Fact 3: One third of the entire world's commerce transits the South China Sea, which is contiguous to mainland China. Fact 4: Any conflict with the PRC over Taiwan or any other Pacific ally will totally disrupt the world's economy, including that of the U.S. If you value your current lifestyle and comforts, you need to recognize this threat. War with China has the potential to change literally everything.

Just how well are we meeting the threat and effectively dealing with it? The media, prominent politicians of both parties, and citizens alike have raised the alarm about our Navy. To assess the Navy, we should look at three main categories: capability, readiness, and leadership.

Capability

In 2016, the Navy said its requirement was for **355** combatant ships. In 2018, the National Defense Authorization Act (NDAA) mandated a Navy of **355** combatants, finally recognizing the increasing threat. Five years later, in 2023, the Navy, as of April 2023, had **296** ships and will actually decrease to **293** by the end of 2024, according to the Congressional Research Service (CRS)[214]. No plan put forth by the Navy or Congress will get us to the mandated number of ships before the 2050s, at the earliest. The Navy and the Marine Corps cannot even agree on the number of amphibious ships we should have. A Congressionally mandated shipbuilding-requirement study, the "National Commission on the Future of the Navy," was created in 2023, but NDAA has not even started to work to pin down an upgraded requirement for ships. That Commission is expected to confirm a requirement substantially larger than the number mandated by the NDAA 5 years ago. Heritage Foundation, in its 2023 Index of Military Readiness Strength Report,[215] concluded that the Navy needs at least 400 manned combatants and assessed the Navy's capability as "weak." Even if the Navy and Congress could agree on the number of ships that are needed to defend the nation against the PRC and others, and Congress actually authorized the money to build those ships, the U.S. lacks the industrial capability to build the ships that we need.[216] Recently departed CNO ADM Gilday said on numerous occasions "the biggest impediment to building the ships we need is lack of industrial capacity."[217] Contrast our ~60 ships in the INDOPACOM area of responsibility with the CCP's 2025 projected 400 combatants, according to the CRS,[218] 150 armed Coast Guard ships,[219] and hundreds more of its Maritime Militia. The imbalance in the Pacific is dangerous to the U.S. and our allies and is getting more lopsided by the day.

Readiness
That same industrial-maintenance capability is likewise woefully under-resourced, as is evidenced by the horrendous record in recent years of maintenance delayed for years and ships taking much longer to repair and at a higher cost than estimates indicated. The GAO reports, in January 2023, that "Sustainment Challenges Have Worsened across the Ship Classes Reviewed." "GAO reviewed key sustainment metrics for 10 ship classes and found that from fiscal years 2011 through 2021, these classes faced persistent and worsening sustainment challenges."[220] GAO reports "From 2017 to 2020, the backlog of restoration and modernization projects at the Navy shipyards has grown by more than $1.6 billion, an increase of 31 percent." And "In 2018, the Navy estimated that it would need to invest about $4 billion in its dry docks to obtain the capacity to perform the 67 availabilities it cannot currently support. This estimate included 14 dry-dock projects planned over [a] 20-year span. However, the Navy's first three dry-dock projects have grown in cost from an estimated $970 million in 2018 to more than $5.1 billion in 2022, an increase of more than 400 percent." Heritage found the Navy to be either weak or very weak in readiness.[221] At a time of increased threat, the Navy has clearly lost ground on keeping ships ready for sea. Unready ships and deferred or canceled maintenance puts an added burden on the ships' crews.

Leadership
Navy leadership has been questioned in very public ways in recent years by an astounding series of incidents, accidents, and sideshows. These leadership lapses have called into question the quality of Navy leadership.

- The collisions of the USS *McCain* and USS *Fitzgerald* led to the unnecessary loss of precious lives. These collisions brought needed scrutiny to the "fighting culture" of Navy's surface warfare community. The attention was so great that Congress members got involved and appointed their own experts to examine the issue in a scathing special report on the Surface Navy.[222]

- The USS *Bonhomme Richard*, a $3B capital ship, burned to a smoking hulk tied up to the pier. The Navy was incapable of fighting the fire, and the ship was a total loss; it was stricken from the records. The JAG manual investigation that followed implicated poor leadership as the culprit, with dozens of senior officers found at fault.[223] That ship was lost to the Navy forever, along with its capability.[224]

- During Covid, there was the incident of the overreaction of the skipper of the USS *Roosevelt* to the pandemic that ultimately evolved into the Secretary of the Navy resigning in embarrassment over his insulting public remarks made to the crew in reference to their former CO.[225]

- A recent survey of Surface Navy officers revealed that most junior surface warfare officers do not aspire to command in their community.[226] How is that even possible that we are recruiting and retaining officers who do not want to be in command? This is institutional weakness personified.

- Two entire ship classes, the Littoral Combat Ship (LCS) and Zumwalt class, are utter failures, and another class, the Ford-class carrier, is still beset with major problems in operational effectiveness more than six years after commissioning. The LCS proved to be so bad and ineffective that the Navy is retiring the entire class decades before its expected lifespan, a spectacular waste of billions. The Zumwalt-class destroyers are likewise an embarrassment, with almost-zero capability after the Navy spent more billions. Instead of being a productive part of the fleet with a proven capability for warfighting, the ship class has been reduced to being multi-billion-dollar test ships. Imagine the embarrassment of putting out to sea with no working weapons systems on board and a Naval gun mount with no shells to fire. That is the Zumwalt class. The

Ford has finally deployed after years of delay, but the Navy admits that it still has not worked out the bugs in five major components of the ship class, most of which impact the ability to generate a high number of aircraft sorties, the preeminent metric for a carrier. And, cost-wise, the Ford class carriers, at $13 B per ship, are costing billions more than was estimated. The CRS has reported in detail about the continuing problems with the Ford class in its regular reports.[227] There appears to be no accountability for these spectacular failures, the wasting of billions, and the lack of adequate numbers of ships for the fleet.

- In the past two years, the Navy has missed its recruiting goals. Even after lowering admission standards, the trend in recruiting has not improved. Fewer and fewer American youth even want to serve in the Navy, and the Navy's message has not resonated with that essential group for the sustainment of the force.

- The Navy continues to have a high suicide rate, despite wringing its collective hands over the problem for years. Covid led to the deaths of 17 Sailors total through 6 December 2022, when the last DOD report was issued.[228] During 2020–2022, the Navy had 194 suicides.[229] More than 10 times as many Sailors took their own lives per year than were lost to Covid through the entire pandemic. Yet the Navy devoted a massive effort to vaccinate the entire Navy, which was always at extremely low risk from Covid. Was there any focused attention to the suicide problem that took dozens of lives each year? There is no evidence to suggest it was treated as any sort of priority. In 2023, the Navy is now trying to recruit enough chaplains so that every destroyer can have a permanent party chaplain onboard. In this author's experience, having a squadron chaplain serving 8 destroyers was effective. Has American culture eroded so much and gotten so fragile that to prevent mass suicides, a chaplain is required on every ship? Is our recruiting screening deficient, such that high numbers of mentally susceptible-to-suicide personnel are being accessed into the Navy?

In no case has Navy leadership risen to the occasion and addressed or solved these seemingly intractable problems. Is that because they are too hard? Or is it the lack of the kind of inspired leadership that was typical of the Navy in the past, from famous Navy leaders of yore like Jones, Barry, Farragut, King, Nimitz, Halsey, Burke, Perry, and Calvert?

Lastly, and the elephant in the room regarding leadership, is the Navy's inexplicable devotion to and zeal in promoting the woke concepts of the political left. Things like Diversity, Equity, and Inclusion, month-long Pride celebrations for the tiny percent of the Navy that are gay, promoting drag-queen shows on ships (the CO who authorized this outrageous conduct by active-duty sailors was selected for flag), accommodating so-called trans-gendered personnel who want to change their gender, and promoting what pronouns must be used are now commonplace throughout the Navy. How do these emphases create the warriors? How do these programs make our ships more lethal? Focus on "social justice" ideology adds not one whit to our effectiveness and lethality as a naval force. There is not a scintilla of evidence that DEI improves the readiness of the Navy. Not explained is how these social fads or influences of our secular society enhance lethality, improve morale or readiness, or make for a more effective Navy to defend our nation. At best they are a harmless distraction and resources drain instead of a laser focus on more important things. At worst, they are an erosion of the core values of merit, honor, courage, and commitment that the Navy emphasized when I served for decades. Those traditional values were what it took to defeat the Soviet Union. And I am proud of what we accomplished bolstered by those values.

Surveying the state of the Navy is a sorrowful exercise for this lifelong Navy man. I grew up in the Navy as the son of a Master Chief, and my entire adult life has been in service to the Navy that I love. Will we have the ships we need to fight our enemies? Will the ships we have be ready for sea? Will the ships be manned by those selected for their high merit and warrior characteristics? Or will our new woke Navy be up to the task even if we overcome the deficiencies in capabilities, readiness, and other leadership failures described above? Citizens must ask themselves and others these important questions. And, if you do not like the answers you are getting, contact your elected representatives, and sound the alarm. Your lives and way of life depend on it. *This article appeared in Real Clear Defense.*

★ ★ ★

Now Hear This, Real Clear Defense, 8 September 2023

America's Moral Compass

By Mike Hollis

As an aviator, I'm a big believer in navigation, and I think America's moral compass needs attention. It just isn't working well, and, in its current condition, it is no longer up to the task of navigating difficult times. Most of our leaders who are entrusted with the task of navigating our country don't seem to be concerned about doing the right thing. They expect us to be satisfied as long as we are comfortable and have growing 401(k) investments. That is not the case.

There is more to America and its citizens than just prosperity. We need to return to being champions of doing the right thing. It's up to individual citizens like us to do that—our government is clearly incapable of acting out of anything other than political motives. In fact, political correctness has actually become a virtue. As a result, we have become accustomed to being lied to in the form of half-truths and assertions that right is wrong and wrong is right. We deserve better than that. Telling bold lies and half-truths to achieve political power is a dramatic failure of the American moral compass. In fact, it is a threat to the U.S. Constitution and the checks and balances it incorporates.

What does "doing the right thing" look like? Look at our history for help with that. America is historically a compassionate and generous country. When disasters befall other countries, we are usually the first to help out. It's in our DNA. We are concerned about individuals and families—not just nations. It should be natural for us to be concerned about the less-fortunate people ("the least of these my brothers" in Bible talk) in our country who are mentally ill, homeless, addicted to alcohol and drugs, and are living in agony. These citizens, many of whom are veterans who have served our country faithfully, need our support for Life Change, which needs to be a bigger priority than Climate Change. Can you imagine what would happen if we committed to fixing these problems rather than "Band-Aiding" them and doing political grandstanding?

It's important to watch how the government is spending our money. You have the right (and obligation) to ask if the government is doing the right thing with our money. It sure doesn't seem like it with a war going on along our Southern border that we aren't bothering to win, in spite of the associated out-of-control smuggling, human trafficking, and slavery. For some reason, we have deliberately caused a monumental humanitarian crisis at the border and have greatly increased the threat to our national security. This was unnecessary and embarrassing—how naive are we?

National security is one of the few things that the government is actually responsible for providing, and our leaders have put us at risk. This is not okay. Politics and greed need to take a back seat to national security. This security concern intersects directly with what we are doing (or not doing) for disadvantaged Americans, who wield little political power. They are being destroyed by the drugs that are flowing across the Southern border and by insane policy decisions. We need to address the related national-security concerns as well as the impact on our individual citizens.

America is still the greatest country on Earth, with a history of doing good, even though some big mistakes have been made along the way. We were not and are not perfect (probably because we are human). We can learn from our history (if it's not erased or altered) and do better.

We can also learn from the history of Israel and other nations. Consider this:

After King Solomon died, Israel was divided into two kingdoms led by his sons. The Northern Kingdom (Israel) and the Southern Kingdom (Judah) were theologically and morally split: Believing Jews who wanted to please God migrated to Judah, while those in the North made Samaria, not Jerusalem, their capital and built a temple there (moral-compass failure). God judged the northern kingdom, and they were killed or exiled by the Assyrians. About 100 years later, Judah also declined morally (another compass failure) and was defeated by the Babylonians. Those who were not killed were exiled into Babylon for 70 years. Afterward, many of the exiles made their way back to Jerusalem (compass repaired and recalibrated after 70 years in exile).

At the time of Christ, they had another compass problem, as the nation's leaders opposed and rejected Jesus as Messiah due to fear of losing political power and privilege. In 70 AD, the Romans leveled Jerusalem and the temple (as Jesus predicted) while putting down a Jewish rebellion. That was the end of the nation of Israel as an autonomous country until 1948. Throughout its history, Israel's national security was tied to the moral state of the nation and the integrity of its leaders. The bottom line is that Israel suffered greatly due to failures of her moral compass and the leaders who were entrusted with the daunting task of navigation during dangerous times.

We should expect no different a pattern or destiny for America. If we do not maintain our moral compass, there will be consequences that affect the health, prosperity, and security of all Americans.

It's time for us to get our compass calibrated and put back into service so we can return to making *doing the right thing* a national priority. This requires us to look objectively at our history and motives so that we can develop and implement a vision for doing the right thing. This is critical and is at the heart of leadership. As Proverbs 29:18 says, *Where there is no vision, the people perish.* Let's not let this happen on our watch!

★ ★ ★

Common Sense

By Tom Burbage

Anyone with a modicum of sense can see there is movement underway to fundamentally weaken the United States as the dominant Republic and the most powerful military in the world. Most concerning is the relentless assault on the most vulnerable (our K-12 children) and the most powerful (our military)—institutions that define who we are. How can anyone deny that there is a slow dismantling of both? When both of those are effectively neutralized, the decomposition of the United States and its overarching Constitution will be complete.

The critical question is how far into that transformation are we? What needs to be done to reverse the decomposition, or is it too late?

Who believes that the critical institutions charged with developing the leaders of the future are immune from this transformation? At what point do parents stand up and demand change? At what point do military leaders "throw their stars on the table" or . . . at what point does the unwillingness to do that become a signal of surrender?

What does the nightly news mean by the Chinese infiltration of Ivy League Universities and the theft of U.S intellectual property and trade secrets? What is the connection between the movement to make the U.S. Service Academies "more like Ivy League universities in their curricula and graduate degrees"? What does the creation of Chinese "affinity groups" at the Service Academies mean? Is there any connection with the top cybersecurity curricula and the foundation of the Hopper Hall Cyber Laboratory? Internet theories gone amok? Maybe a year ago, but certainly questionable now.

What does the new focus on "Diversity, Equity and Inclusion" really mean? The Superintendent Letter to the Brigade in September 2020 indicated a need to align the Brigade with his vision of where the Naval Academy should be headed. In the interim, an office of DEI and an infrastructure of "Diversity Peer Educators" or DPEs have been installed. The basic Leadership Pillar of the Naval Academy that flows from the Commandant through the Battalion and Company Commander/Officer now has a separate channel for resolving "DEI Issues." How do these insidious changes on the margins, taken singly as insignificant, in fact, taken in the aggregate, support the concept of radical change and create a time-sensitive doubt in the future of our country?

So, why are we concerned, and where do we go from here? It is almost as if the Founding Fathers and everyone who has served and died in defense of our Constitution has suddenly become irrelevant. Despite the fact that all of our leaders take an oath to defend and protect it against foreign and domestic enemies, it is not happening today. The Obama regime, despite its apparent focus on diversity as a clear signal that we, as a nation, were moving into a new generation of inclusion, took a hard left turn and began the generational drift to the current situation we find our nation in. The slow purge of warrior leaders in the military was the beginning but was not visible to most of the nation. Trust but don't verify was a serious fault in our democratic DNA against domestic assailants despite President Reagan's caution with foreign enemies. We, as a nation, forgot the fact that our Constitution clearly warns of domestic enemies as equally powerful as foreign destructive forces to our future. The domestic enemies have clearly arrived, and we seem to be powerless to rebuke them.

So, what do we do?

1. In the immediate, parents must continue to challenge the school boards and recapture the educational processes of our children.

2. Alumni of institutions that contribute the funding that is the lifeblood of their alma maters need to seriously challenge the future objectives of their universities.

3. In the long term, there must be an awakening to the reality of the cliff we find ourselves on as a nation. We must make sure that every

vote counts and that the ability to wrongly influence our election process is fixed once and for all. The 2020 election will someday be a Harvard Business School Case study in the ability of criminal elements to change the outcome of a democratic process.

4. Make our Congressional supporters defend and protect the leadership-development process for the future military, government, and industry leaders who come through the Service Academy process. This requires a focus on development of warrior leaders and a return to a meritocracy-based, diversity-sensitive approach. This should be the focus of our 2024 actions.

★ ★ ★

America Needs Leaders with Courage

By Randy Arrington

When I taught political science at UCLA, I would begin by postulating this humble premise to my students. Politics is simple. Politics is easy. In fact, the undergirding dynamic of politics hasn't changed since Plato and Aristotle were first conjuring up their political theories some 2400 years ago in Ancient Greece.

Politics is the authoritative allocation of value in society. Simply stated, politics is about Who Gets What, When, and How. Value and values are disseminated from the myriad of resources we possess as a nation, both tangible and moral.

Certainly, our national assets are indeed abundant, but they can be needlessly squandered into extinction. There is a limit to what we can do with our national treasures.

Contemporary politics has evolved into a complex leviathan because We the People have allowed politicians, activists, and university professors to deceive us into believing that politics is extremely complicated. And, as such, it requires their particular brand of genius to comprehend and properly manage all of the intricate variables so that our society will be successful and enjoy longevity.

Power and prestige have indeed corrupted our politicians and our academics during this evolutionary process. There is no worse heresy than the office sanctifies the holder of it. But, again, We the People must share the blame because we are not electing our best and brightest citizens, as James Madison predicted we would, in *The Federalist Papers*.

The Founding Fathers created America as a nation based upon Locke's notion of consent of the governed (not divine right), and upon limited government intrusion to maintain order and liberty. Government was to protect the nation against invasion, both foreign and domestic, and to create an atmosphere of liberty. Freedom would encourage people to pursue opportunities to succeed in life, according to individual desires.

As long as citizens didn't harm others during their pursuit of success and happiness, the government apparatus would not be an impediment to them. In other words, citizens' liberty would not be removed or incrementally chipped away by the government. Remember this: Liberty is not the power of doing what we like, but the Right of being able to do what we ought to be doing.

Although Order, Freedom, and Equality are the political goals that our American government pursues today, Equality in a social context didn't become a formally recognized objective in American society until Lincoln's Emancipation Proclamation. Economic Equality as a political value was solidified in the 1930s, when Roosevelt's socialist public-policy agenda was adopted.

In a society, truth often changes because truth is routinely subjected to human interpretation, propaganda, and newly concocted circumstances. What is highly valued in one generation can easily become the pariah of another. What is truth in one generation can easily become a lie in the hands of another.

Here is a good example: Hard work, ingenuity, and determination are the pathway to success in America. Americans of our Greatest Generation understood this truth. They would literally pull themselves up by their bootstraps and work extra hard to become successful, despite the many obstacles thrown into their path.

This eternal truth slowly became distorted in the minds of progressives, liberal academics, and socialist politicians, who successfully brainwashed activists, students, and constituents into believing it to be a lie.

Today many Americans believe that no amount of hard work will make them successful because the deck is unfairly stacked against their success. In fact, many believe that their continued failure in life is the direct result of a conscious conspiracy by some specified group of selfish, enterprising people. Therefore, the remedy for this unfair situation is that government must fully subsidize the failed citizen's existence and forcibly deny liberty to those fortunate enough to be successful.

The express purpose of socialists is to redistribute success and value to those among us who are allegedly less fortunate because of circumstances beyond their control.

In the 21st century, we do not have Majority Rule in America, despite what we were all taught in the fifth grade. Today, small cadres of people often dictate how the majority of citizens can and will live in society.

Professor Brody at UCLA taught me that, in America, we have Minority Rule. As such, we must continually be on guard because the "Tyranny of the Minority" is much worse and more damaging than the Tyranny of the Majority. Tyranny of the Minority has the potential to permanently alter the way people actually think and analyze the political context in which they live.

The Founding Fathers envisioned Congress as public service; it was never intended to be a permanent, lifelong career as it is today. Remember this, where you have a concentration of power in a few hands, all too frequently men with the mentality of gangsters get control. History has proven that all power corrupts; absolute power corrupts absolutely.

America needs strong leaders who will stand up and boldly speak the truth, despite the media onslaught and vitriol that will be hurled at them on a daily basis. This is a daunting task to ask of any potential statesman, but our nation is worth such personal sacrifice. The danger is not that one class is unfit to govern. Every class is unfit to govern.

Great men are almost always bad men. The individuals that we elect must surround themselves with staffers who keep the officeholder firmly grounded in humility, always reminding him or her that they are public servants, not rock-star celebrities. Elected officials must never forget that the people they serve actually hold the power.

Elected representatives merely exercise political power on behalf of their constituents. We all need to be students of history because history is not a burden on the memory; it is an illumination of the soul.

Let Freedom Ring
God Help Us

★ ★ ★

"**America Needs Leaders With Courage**"—This article was first published in Politics in America: Lecture Notes of a Lunatic Professor Volume 1 (IUniverse 18 February 2016)

You Decide

By Tom Burbage

Words matter. One of the historical words that has always mattered in the political world is "Reconciliation." According to *Webster*, an act of *reconciling* occurs when former enemies agree to an amicable truce. But there are a couple of other words in that definition that need to be unpacked. The words "former" and "amicable" and "truce" need some additional clarification. "Former" generally means "traditional," and "enemy" is generally regarded as a *foreign* enemy—another nation seeking to disrupt our democracy and our American way of life. After all, there are many nations on this planet that do not enjoy our freedoms and our liberties. But those qualities are precious enough that there was a time when we could recruit and mobilize our youth to fight and die to sustain that vision of our forefathers.

We've been well prepared and well invested to defend against foreign enemies, but what if it's not a foreign invader? Our forefathers warned us about that other enemy, the domestic one.

Ronald Reagan warned us: "Freedom is never more than one generation away from extinction. We didn't pass it to our children in the bloodstream. It must be fought for, protected, and handed on to them to do the same, or one day we will spend our sunset years telling our children and our children's children what it was once like in the United States, when men were free."[230]

I am now in the sunset of my years, and I am preparing my speech for my children and grandchildren. The domestic enemy has infiltrated our Republic, and it is well on its way to changing the freedoms and our liberties that our children and grandchildren should be inheriting from us.

Have we reconciled our differences with foreign enemies? The disastrous withdrawal from Afghanistan, the tolerance of Chinese balloons over our sovereign territory, the apparent ignorance of the proliferation of Iranian drone warfare, and the overarching policy of appeasement would indicate we have not. Depleting our weapon stockpile in the proxy war with Ukraine while simultaneously depleting our strategic petroleum reserve (critical to any logistics support in the event we actually may have to fight) both imply that our foreign enemies are not a priority.

We also face an unprecedented deficit in recruiting. Loss of patriotism driven by the focus on race-based conflict in our public-school systems and universities has affected our ability to recruit, equip, and train the next generation of freedom fighters. Coupled with the impact of forced social experimentation in the form of Diversity, Equity, and Inclusion on our military-force structure, recruiting is exacerbated by early exits. The new answer seems based on the lowering of standards to maintain numbers.

Consider the occasional Presidential speech called the "State of the Union" address—we should all be very concerned that we do not have a mutual understanding of that State. It should be measured, somehow, on the health of our Democracy, which has always been the metric that has separated the United States from the rest of the world.

Let's list a few concerns by age of our population:

Public Schools: the source of recruiting for our military and the relative ranking of the product of our schools measured against the school systems of our global competitors. How are we doing? Not well. Consider the corrosive influx of Wokeness and its basis, Critical Race Theory, Transgenderism, and revisionist history. Thankfully, the pandemic allowed parents to see what their children were being taught, and a vocal rejection began. "Division" used to be a math term, but it is now an ideology term.

Teenage (and adult) years: Open borders have opened a flood of Fentanyl, which has created an epidemic affecting multiple generations as young as school children. The complete disregard for the influx of millions of refugees and the impact on our society, especially the border states, is profound. The free cartel operations and the influx of deadly drugs are a scourge on our Democratic principles.

Adults: The complete abandonment of energy independence, ignorance of the reality of our tax system (Tax the rich?), unbridled wasteful spending and the associated misuse of federal tax dollars—under very misleading labels—and its accompanying inflationary effects on the middle and lower segments of our society are conveniently ignored. Our society has lost its moral compass. Only witness the recent Grammy Awards to see the depravity on prime time.

Hearkening back to our childhoods, schools issue report cards... so how do we initiate a report card for our leaders, who seem to act like unaware and uninformed children? You decide.

How do we fight a domestic cultural war we never wanted and never thought we would fight? How could a strong Democracy allow such a war to start? You decide.

Once you decide, you have two choices. You can act, or you can watch. The first might make a difference, the second will not. You decide.

★ ★ ★

You Decide, Armed Forces Press, 10 February 2023

Diversity ... What Can Be Learned from Fruit Bowl Preparation

By Bruce Davey

If I am putting together a bowl of fruit, I might very well think about having some diversity in the content. The beauty of the presentation may be of real significance, depending on the function of the bowl of fruit, and the superficial colorings of the individual fruits need to be taken into account. The arrangement of the various types of fruit would be of significance also if how the bowl looks is of primary importance, since the shape of the fruit will need to be taken into account. Lastly, the coloring of the fruits involved would dictate their position within my display, and I would consider the surroundings within which the bowl will sit.

All of this is true if the mission of my bowl of fruit is to be decoration for a setting and that this mission is solid and unchanging. I am not expecting my bowl of fruit to be used for any other purpose, and the diverse nature of its contents is expected to add to the beauty of the milieu, adding demonstrably to the setting only in color and presentation. Diversity here is of huge significance, as anyone who compares a mixed-fruit bowl with a box of oranges can easily discern. The mission is to be visually pleasing, and even emotionally strengthening, in the relaxed atmosphere of dining. Great mission definition, solid mission planning, excellent mission accomplishment.

Unfortunately, even in our banal example, things can go awry. What if the beauty mission assigned to the bowl of fruit is overwhelmed by a need for nutrition and subject to the whims of availability? What if the bowl of fruit is now serving as a source of sustainment for a number of hungry immigrant families who have no interest in beauty and are desperately in need of nourishment? The mission has now changed, and the performance of the bowl arranger should accommodate such a change in mission. I must now find a bowl filling that is at once easy to obtain and carries the largest nutritional value. Outward beauty, coloring, shape, and size are not within the selection criteria provided that they fit into the bowl. I am utterly subservient to the overwhelming dictates of the mission.

From the curious example of the bowl on a table, one may reach the conclusion that mission definition is the key element in bowl-content planning and accomplishment. Not having a career in bowl decoration does not preclude the application of these principles in one's occupational pursuits.

Warriors cannot but agree. There may have been a time in the U.S. military when our strength and ability were such that we could focus on matters which seem somewhat far afield of our primary purpose of protecting our nation from our enemies. When you are throwing 90 mph heat in a Little League game, there is no real reason to worry about your third baseman's fielding abilities, and you are at liberty to ponder the dugout's décor. We are no longer living in that pleasant circumstance. Our preeminent position within the roll of nations has become tattered by any number of poor decisions, bad leadership, loss of focus, and poor selection of those to whom mission definition, planning, and accomplishment have been entrusted. America is under attack in every sphere of life. Our financial, educational, moral, and constitutional structures are weakening, and our enemies—even our friends—recognize this fraying. The importance of maintaining the Common Defense structure, the last redoubt, in a robust fashion is not just important—it is of primary importance. Within that group which has been charged with such maintenance, there needs to be an urgent understanding of mission. That mission is not to overcome cultural difficulties many years in the making. It is not to address difficulties which exist within our society for certain groups of people who are or have been underrepresented. It is not to use any metric for decisions outside of furthering our nation's safety. If programs, plans, training, and selection have any targets which cannot be clearly and definitively shown to further that mission, they should be scrapped immediately.

As a Navy fighter pilot, the father of a Navy fighter pilot, and having had a career in the operation of aircraft, I can say definitively that the racial, skin color, sexual preference, and religious makeup of those with whom I have been in contact over the last sixty years has had absolutely nothing to do with how they were able to perform the missions to which we were tasked. It was the training of excellence, the determination of the individual, and persistence of the effort that made them capable as warriors. And these are the attributes for which the military has looked, not the color of the skin or the societal background of the applicant.

We warriors have accepted diversity as it presented itself within the boundaries of meritocratic advance. It has not been a target; it has not been a mission of our military. That falls within the purview of society as a whole, as the society arrives at its mission definition. But, like the bowl of fruit, our military must be planned and operated with an understanding of its mission. It is not to address historical inequities, but rather only to defend our nation. The leadership of our Armed Forces must understand that the mission that they are vested with is that and only that. The hungry man presented with a beautiful bowl of fruit that is made of wax is destined for a period of hunger due to the lack of this peculiar understanding of mission. Our prayer should be that our military and civilian leadership will regain their understanding of the mission of our Armed Forces. Just as in our personal life, we would not care how diversified the group is that came to rescue us from a burning building, but rather how well they can fight the fire and carry the injured. Mission definition and understanding are critically important now in America's Defense Department leadership, and they are presently not at all in evidence.

★ ★ ★

The Navy: Dead in the Water?

By Brent Ramsey

"Mission: The United States is a maritime nation, and the U.S. Navy protects America at sea. Alongside our allies and partners, we defend freedom, preserve economic prosperity, and keep the seas open and free. Our nation is engaged in long-term competition. To defend American interests around the globe, the U.S. Navy must remain prepared to execute our timeless role, as directed by Congress and the President." The preceding statement is from the U.S. Navy's website.

There are many indicators that the Navy is at increasing risk of mission failure.

1. Missing recruiting goals by thousands for two years in a row; missing its goal for FY 2023 by more than 7,000 new recruits. The impact of missing recruiting goals is cumulative. Its impact does not subside if, in subsequent years, deficits are not made up. Lack of manpower adds to the strain of a Navy struggling to meet its national priorities overseas. Failing to recruit enough people to man the Navy is a result of many factors. Since the Afghanistan debacle, the public's faith in the military has plummeted to new lows. With relatively low unemployment, the competition for young people is high. American youth are less fit, less capable of serving in the military than at any time in our history. Fewer young people want to serve, as the political left teaches them to hate our country, academia promotes socialism, and race hustlers malign our country for its supposed racism and White

supremacy. Divisive ideologies like Critical Race Theory and Diversity, Equity, and Inclusion are now promoted vigorously up and down the chain of command in the Navy. These ideologies alienate the youth of what for generations was the most fertile recruiting grounds—White, southern, Christian Americans. This demographic is now increasingly averse to serving in our new politically correct Navy of DEI, Pride month, correct pronouns, drag queens, and transgender people. If the Navy cannot recruit now for the existing numbers of ships we have, we have no hope whatsoever of filling out the ranks of a Navy with a much higher number of ships.

2. Recently, due to the international wars simultaneously in Ukraine and Israel, and high tension in the Taiwan Strait/South China Sea, the U.S. Navy had an almost unprecedented eight Carriers at sea at the same time. The only three not at sea were unavailable due to long-term maintenance. Normally, the Navy might have three or four carriers at sea at one time. Navy ships and crews continually operating wear out rapidly. Typical deployments last six months. The *USS Ford* has been deployed for seven months, and SECDEF just extended its deployment in the eastern Med for the second time. The longer the deployment, the more worn out the crew, and the higher the rate of equipment failures become. As deployments go on for longer and longer, the size of the crew shrinks due to illness, pregnancy, injury, and suicides. Typically, ships returning to home port after a lengthy deployment are missing a substantial number of the deploying crew. This puts much more stress and strain on the remaining crewmen. The international situation, with multiple wars demanding our attention simultaneously, is eroding our Navy's readiness at a high rate. When the ships and their crews wear out, there will be no alternative but to return them to port for re-fit and rest for the crews, regardless of whatever pressing mission the ship is on. That the Navy does not have enough ships is now obvious to even the most casual observer when multiple hotspots in distant seas occur. When the proverbial stuff hits the fan, the very first question everyone, including the President, asks is, "Where is the nearest carrier?"

3. The Navy's high suicide rate over a lengthy period demonstrates the leadership's tragically being unable to ameliorate the problem. The higher the OP tempo, the longer the deployments, the more arduous the maintenance periods are, the more inadequate berthing arrangements are for ships in long-term overhaul; this, in turn, aggravates already-high-stress environments and seemingly makes things unbearable for too many of our Sailors. The Navy seems content to muddle along with scores of Sailors killing themselves year after year and the heart-rending loss of life continuing as an unsolved problem. We Navy folk like to call ourselves "warriors," and most of us fit the description of selflessly putting ourselves in harm's way for the benefit of others, for the benefit of our nation. But what does it say about our culture to have so many warriors who end their own lives because somehow our organization does not recognize their despair until it is too late, and they have taken the irreversible step and ended their own life? Considering how extremely selective the Navy is at screening those who volunteer to serve, why do such high numbers of exceptional citizens, with all that the Navy has to offer, choose to end their own lives? Are our leaders so overwhelmed by the work the Navy has them do that they cannot be close enough to their Sailors to recognize those who are *in extremis* in time to help them?

4. Notable institutional leadership failures in multiple major-program areas and multiple high-profile operational failures are now far too common. Examples include well-documented cases such as the LCS and Zumwalt ship classes, the *USS Ford* class's cost overruns, lateness, and multiple ship systems not being fully operational (EMALS, ammo elevators, arresting gear, etc.), even years after being in commission. An egregious example of a mammoth leadership failure was the loss of the *USS Bonhomme Richard*, a multi-billion-dollar capital ship that, due to negligence, was allowed to burn at the side of a pier, a $3B loss with no replacement. A total of 45 Navy leaders were disciplined due to this one incident. The grounding of the *USS Connecticut* resulted in this vital attack submarine being out of commission for years for repairs. The *USS Gettysburg* has been out of commission for

more than eight years undergoing modernization. Four of the seven cruisers selected for modernization will instead be decommissioned after the Navy spent billions on upgrades. The collisions of the USS *McCain* and USS *Fitzgerald* with commercial shipping were failures of leadership that led to the deaths of 17 Sailors.

5. In the 2018 National Defense Authorization Act, the Congress established the size of the Navy to be 355 battle-force ships. According to the United States Naval Institute as of 6 November 2023, there are currently 291 battle-force ships in the Navy. The predictions from the Congressional Research Service are that the size of the Navy will stay relatively the same for the rest of this decade before it slowly starts to increase in size in the 2030s. In 2022, then-CNO Gilday announced that the requirement is actually much higher, in excess of 500 battle-force ships. Multiple other experts' analyses confirm those higher numbers. The PRC's PLAN (People's Liberation Army Navy) is already at 350 combatants and building at a rate that is, at least, four times faster than that of the U.S.

6. In the FY 2023 NDAA, there was a provision to establish a Commission to study the Navy and its requirements. The report of the Commission is due to the Congress by July 1, 2024. As of this writing, the commission has not even been formed. The Secretary of the Navy and the CNO should be urgently pressing Congress to get this Commission up and running. Furthermore, the Navy should be proactive in suggesting that Navy advocates serve on the Commission or serve on the staff of the Commission. It is vital for the defense of the nation to have the definitive knowledge of what the Navy's true requirements are in 2023 in the face of multiplying threats all over the world.

Conclusion: All of these factors outlined above make it clear that our Navy is *in extremis*. There are not enough ships to do the mission or enough manpower to man the ships optimally. Deployments are too long, and our people and ships are wearing out. Recruiting is stagnant. Too few ships, not enough people, not enough shipbuilding, or low repair capacity have us on the brink of mission failure. To put the size of the Navy in perspective, when this officer went aboard ship in 1970 to conduct anti-submarine patrols, looking for Soviet ballistic missiles submarines, the Navy had 792 battle-force ships in commission. We now have 291. Then we had a Cold War against one adversary, the old Soviet Union. Today we have adversaries all over the world and are trying to perform the mission quoted above with a tiny fraction of the ships we had decades ago. As a maritime nation with treaty allies all over the world, coupled with our dependence upon the sea for 90% of the commerce that keeps our economy running, it is a travesty that such neglect of the Navy has occurred. Who is at fault for this neglect? Congress is ultimately at fault, as it holds the power of the purse. However, it is incumbent upon senior Navy leaders to make the case for the right-size Navy. The CNO and every other Navy flag who testifies before Congress should be sounding the alarm about the imminent failure of the Navy to perform its mission now in "peacetime," with multiple hotspots in Europe, the Middle East, and in the Taiwan Strait and South China Sea, and, even more importantly, in the next actual fighting war. Someone long since should have laid his stars on the table to make the point to politicians that we need more ships and more manpower for the survival of our nation. Our way of life and our very lives are at stake if we do not rebuild our Navy to an adequate size to perform its vital worldwide mission. *The article originally appeared in Real Clear Defense.*

★ ★ ★

The Navy—Dead in the Water? Real Clear Defense, 5 December 2023

He Who Will Not Risk, Cannot Win

By Tom Klocek

Along with "Don't give up the ship," these words of John Paul Jones, one of the greatest Naval heroes of all time, have guided our Naval warriors and leaders since the Navy's inception in 1775. John Paul Jones was not only a great seaman and Naval warrior; he also understood leadership and what it takes for success in battle, particularly at sea. He understood the importance of those who made up the crews of a Naval ship as evidenced by another of his sayings, "Men mean more than guns in the rating of a ship." There is no sense of "Diversity, Equity, and Inclusion" (DEI) in any of those statements. I sincerely doubt that he would agree with our current Chief of Naval Operations, who said recently, "Our strength is in our diversity."

When John Paul Jones wrote about the "Qualifications of a Naval Officer," he said nothing about race, or equitable outcomes, but only about the merits of an officer, including such concepts as gentlemanliness, charity, and honor. The whole tone of the letter shows the need for personal responsibility. He also encouraged rewarding meritorious actions while not overlooking faults. Nothing in Jones's writing or in the history of the Navy in all of the wars since then addressed "diversity." While efforts and programs existed to facilitate equality of opportunity regardless of background, ethnicity, or religion, final outcomes were based on performance, not quotas. I doubt that John Paul Jones would even consider a need for safety and inclusion. Remember, this is the man who said, *"I wish to have no connection with any ship that does not sail fast, for I intend to go in harm's way."* That doesn't sound like someone looking for a "safe space." These foundational values of the naval officer and the oath of fidelity to the Constitution and its laws, among them the Civil Rights Act of 1964, obviate the need for anything more. Those who fail to embrace and follow them should be discharged from our Navy.

The USNA DEI Master Plan mission statement, which begins, "**Attract, retain, and develop a diverse cadre . . .**" is missing a key word: "capable." It has, as objective one, to "*Become a school of choice because of an inviting, safe, and supportive campus where everyone feels they belong and have equitable opportunity for success—regardless of race, ethnicity, culture, gender or socioeconomic background.*" The elimination of merit as a primary consideration in recruiting, training, and retaining midshipmen is an obvious weakness in the plan. As I recall my days at USNA and in the Navy, one had to earn the support of others. One was given a "grace period," during which they had the opportunity to prove themselves, show their intent and ability to become a member of the team, and be willing to embrace the goals and mission of the unit (squad/company within the class at USNA, division/department/ship at sea). If they failed, they were "washed out." There were no "safe spaces" to which you could run, although often a new member of the team would find, at least for a minimal time period, a mentor in whom they could confide, discuss issues and concerns, and obtain guidance. This fostered the development of both individuals as well as building understanding and teamwork in the subordinate.

The above master-plan statement appears to ignore the concept of adversity in the development of leaders and teams. Having had an older brother who graduated three weeks prior to my entering the academy, I ran into several upperclassmen who "had it in for me" (my brother wasn't popular with everyone). I had to deal with being singled out for special attention, and it only made me more determined to get through Plebe Year. When the brigade returned the next fall, one of those upperclassmen, on passing me in the corridor, congratulated my tenacity and determination, saying, "Glad to see you made it." As the philosopher Friedrich Nietzsche once said, "That which does not kill me makes me stronger."

When I was executive officer of a guided-missile destroyer, I personally greeted and interviewed every new addition to the crew. I impressed upon them that each man was an important part of the crew regardless of his position. I pointed out that, at some time, without knowing it, the lives of some or all shipmates could be in their hands, whether they were the officer of the deck, a lookout, or a food-service worker. In this way, I was trying to emphasize not only the importance and significance of each individual member of the crew but also the need for them to take responsibility for their actions. Was this effective? Did it foster team building and pride in being members of a cohesive fighting unit? I believe so. I base this on the fact that, during a port visit while deployed to the Mediterranean Sea, some of our crew met up with crewmembers of another ship. These other crewmembers were bragging about how easy it was to get around the rules and the leadership, and to obtain drugs on the ship. Our crewmembers responded to them that "we don't do that on our ship." Later in the cruise, a few ships were selected by the fleet commander to visit a port at which the fleet was refused visiting rights because of past negative behavior. Our ship was one of the four selected to visit. We had a good reputation. (Later in history, both of the commanding officers I worked with went on to achieve four-star rank and serve as Vice-Chiefs of Naval Operations. Good, meritorious leadership has its rewards.)

The Navy has had a long history of holding people accountable at all levels of command. It has always been important that decisions be made at the lowest level possible, consistent with the rules and regulations, good engineering and seamanship practices, and capability (merit/leadership). Establishing separate DEI chains of command goes outside good leadership and is divisive in and of itself. It is akin to the policy of having commissar overseers once found in the Soviet Navy, which weakened initiative and leadership and instilled timidity rather than boldness of action. A far cry from intending to go in harm's way.

Furthermore, sadly, increasing diversity usually means a reduction of standards. There is a reason for the old adage, "A chain is only as strong as its weakest link." Do we want to recruit people who can do the job, or do we just want people who meet some mythical breakdown of ethnicity, faith, sex, or whatever, without regard for merit? Already the services are relaxing recruitment standards to increase the availability pool. This relaxation includes relaxing age and qualification-score requirements. In this treacherous international environment, weakening the chain is not good policy.

The Navy's embrace of DEI is the wrong path; it will erode quality and damage our ability to accomplish the mission. Already we are seeing retention and recruiting challenges that have diminished the Navy's effectiveness across the board. The public's trust in the military in recent years has plummeted to the lowest level ever, partly due to its politicization. Navy leaders must reject politics and focus on our bedrock core values as espoused by John Paul Jones all the way back to our founding. Nothing else matters.

This article originally appeared at AmericanThinker.com[231]

★ ★ ★

"**He Who Will Not Risk Cannot Win**" originally published at AmericanThinker.com on 26 January 2023 under the title "Who Will Not Risk Cannot Win"

A Salute to Our Veterans and a Shared Concern

By Tom Burbage

All Americans should salute our veterans, who have given so much in defense of the freedoms we enjoy as a unique nation in the world. At this time of reflection on their sacrifices, there is a movement afoot that represents three perspectives. See if you share this view.

1. As Americans, we are concerned about the growing crisis in our nation, our society, and our institutions. Over the last 80 years, America has lost hundreds of thousands of lives and spent trillions of dollars from our national treasury fighting against totalitarian governments. We now seem to be on a path that would allow the same divisive ideology to find a home inside our nation. There seems to be a slow awakening to how deeply it has become engrained in our institutions of higher learning, in many of our federal-government organizations, and in corporate America.

    Most concerning to us is that it is now infecting the two institutions which should be most protected: K-12 (*the impressionable, not fully developed minds of young children*) and the military (*whose members write us a blank check of service with their lives as collateral*). This should be a major concern to all Americans.

2. As veterans, we are concerned about the impact of the changing ideology on the erosion of military readiness and combat capability. Some view the military as the perfect Petri dish for experimentation: a

very large, captive sample set, comprising a diverse slice of American society, and subject to government edict and funding.

But veterans all took an oath to uphold and protect the Constitution of the United States against all enemies, foreign and domestic. All took their oath as a lifetime commitment. For the first time, we are sensing a real domestic enemy emerging.

It is also a matter of **Stakes and Mistakes**: We believe the **stakes** are high and getting higher as real enemies see the weaknesses in our current leadership. Taiwan and Afghanistan redux are only two; there are several others. Military readiness, not political positioning, must be prioritized. The recent **mistakes** (surface and subsurface collisions at sea, loss of a new capital ship through arson, withdrawal from Afghanistan [the loss of extensive military equipment to a terrorist regime and the abandonment of Americans and Afghani-American supporters]) would point to a loss of focus within our military command structure.

3. As alumni of our Service Academies, we are concerned about the impact of these changes on the unique role that the Service Academies play in transitioning the best and brightest of our youth to the future military, government, and industry leadership. Service Academies must deal with incoming high-school students, in some cases indoctrinated with the DEI/CRT culture and often with very different family values than our older generation would recognize today. At the same time, the institutions must deal with the flow-down of social experimentation from left-leaning administration policies. While we all believe in and support the Mission of our Service Academies, we feel an obligation to vocally verify that maintaining a warrior culture as the basic element is critical, not adapting these unique institutions to the whims of societal change.

The Bottom Line

The American people must awaken to the real domestic enemy that now threatens our American way of life. Somehow, we have allowed it to happen; have we been asleep at the switch? No, actually, in the words of Ronald

Reagan, we trusted leadership to make the right decisions, but we failed to verify, one of the DNA shortcomings of a Democracy. We assume leadership of our critical institutions have the best interests of our Constitution and Democracy in mind. That has proven to be a bad assumption. There has been a slow awakening to this threat, both in our schools and in our military. Our democratic historical analogy is the unleashing of the sleeping bear. It is time for the bear to come out of his hibernation.

★ ★ ★

There Is No Justification for the Navy's DEI Push

By Brent Ramsey

The Constitution is the supreme law of the land. The oath that I took upon commissioning was a solemn vow of allegiance to that Constitution and says, in part: "I do solemnly swear that I will support and defend the Constitution of the United States against all enemies, foreign and domestic; that I will bear true faith and allegiance to the same . . . so help me God."

That oath that I took 54 years ago, when I joined the Navy, continues to bind me to my country.

Other core Navy documents show the military servicemembers' oath is not limited to the plain text of the Constitution, but to its spirit and intent as well. The Sailor's creed, for example, affirms that Sailors are "committed to excellence and the fair treatment of all." One of the Navy's core values holds that the "day-to-day duty of every man and woman in the Department of the Navy is to join together as a team to improve the quality of our work, our people, and ourselves."

These fundamental values and truths obviate the need for introducing divisive social-justice ideas into our Navy. Yet that is exactly what the leftist bureaucrats and military leaders who run our Armed Forces have done.

Since 2011, the Navy has been ordered by former President Barack Obama and President Joe Biden to implement Diversity, Equity, and Inclusion. Rather than rejecting this leftist initiative as antithetical to the values of honor, integrity, teamwork, and ethical behavior that the Navy has sworn to uphold, the Navy has completely embraced divisive DEI and implemented it throughout the force.

There is no justification for the woke takeover of the military. Research shows no connection between diversity based on skin color and excellence in military and/or tactical performance. In fact, Task Force One Navy had a charge to discover the connection between diversity and excellence in tactical performance and failed to find it.

Moreover, the Navy is already incredibly diverse. After the Defense Department-wide stand-down in 2021, a survey was done to assess so-called "extremism" in the military. A senior official told me: "I reviewed the stand-down brief that was internal to Defense Human Resources Activity, and it included survey results that indicated, overall, that 2% of DOD personnel are concerned about hate crimes or racism."

Why are millions of dollars and hundreds of thousands of hours being devoted to implementing DEI in the military when there is no justification or appetite for it?

It is time for the senior leaders in the Navy, the admirals, to wake up to the fact that DEI programs mandated by the executive are unnecessary and harmful to the Navy. There is no evidence the Navy is systemically racist, and pretending that it is harms morale, recruiting, retention, and readiness. *Originally appeared in the Washington Examiner.*

★ ★ ★

There Is No Justification for the Navy's DEI Push, Washington Examiner, June 2023

Liberty Matters More

By Jim Tulley

During the Second Virginia Convention on March 23, 1775, at St. John's Church in Richmond, Virginia, Patrick Henry gave a speech in which he said, "Give me liberty or give me death." History has recorded countless patriots who have given their lives for the preservation of liberty. While the final words of our national anthem proclaim, "The land of the free and the home of the brave," it has also been said that we are, "The land of the free because of the brave."

We continue to ask ourselves, what is it about freedom and liberty that makes citizens of a free society willing to die for it? It happens all over the world, be it Hong Kong, Tiananmen Square, 1956 Hungary, colonial India, and now even Cuba. What led these people and many others during such historic times and events, to be willing to die for freedom? The only logical answer is that such feelings are enshrined in the human spirit. The only logical answer is that Thomas Jefferson was right: "We hold these truths to be self-evident, ... that we are endowed by our Creator with certain unalienable Rights, that among these are Life, Liberty and the pursuit of Happiness."

Now we are engaged in a war of words with two prominent, Marxist-based philosophies, Black Lives Matter (BLM) and Critical Race Theory (CRT). BLM was founded by two self-proclaimed Marxists and has among its seven demands, to "launch a full investigation into the ties between white supremacy and the Capitol Police, law enforcement, and the military." Of all the institutions in America, why has BLM picked out the military for investigation or worse? Might it be because our military and the veterans who served are the last line of defense in a cultural takeover of this country? If our military becomes just one more social-justice organization, then who will join? Who will be willing to put their lives on the line for freedom and liberty?

The BLM website states, "Black Lives Matter Global Network Foundation, Inc. is a global organization in the US, UK, and Canada, whose mission is to eradicate white supremacy and build local power to intervene in violence inflicted on Black communities by the state and vigilantes."[232] *Violence inflicted on black communities*—really? Numerous commentators have called attention to the fact that BLM has shown no concern for black-on-black crime in our inner cities, where violence has become the norm.

On Feb 6, 2021, a *New York Post* editorial stated, "Last year saw the largest year-to-year increase in homicides ever recorded in U.S. history ... Victims of these homicides are disproportionately African American. At least 8,600 Black lives were lost to homicide in 2020, an increase of more than 1,000 compared to 2019."[233] Do these lives not matter, and when will we see a BLM investigation related to how to reduce such violence that is plaguing our cities?

I do not recall hearing of a single person who stood up and said he is willing to give his life for Black Lives Matter. Most of what we hear is about their desire to take lives, especially White lives, in the cause of Marxism. Where is the heroism in such a philosophy, and is there anything for which they are willing to give their lives?

To make matters worse, we now see CRT and its "woke" philosophy infiltrating our educational institutions and our society. There can be little doubt that CRT has its origins in Marxism, and it is now metastasizing through our culture, even into the military and all three Service Academies. If this continues, how will future leaders be adequately trained to defend America, when CRT is telling them that we are an intrinsically White-supremacist country, worthy only of hate and condemnation?

CRT is even more dangerous because it can be far less obvious. It is a covert tactic with the same freedom-destructive strategy. It is an insidious philosophy which countless educators have insisted they do not teach at their institution. They are usually correct in that nowhere have we found a CRT-101 course in any curriculum. The other often-proposed argument is that, in any military organization, it is important to understand the methodology of the enemy. With that we would agree, but there is still a big difference between teaching and preaching. CRT is a "Virus of the Mind" (Richard Brodie book)[234] that can infect without recognition.

BLM and CRT comprise a double-edged sword that will destroy us from within if we allow it. As both are related to Marxism, it is important to have a clear picture of how and why they have the effects we see. One might say that BLM is the hammer that pounds our compassionate sensibilities, because who can be against lives that matter? The argument is compelling and memorable. Meanwhile, CRT is the sickle that cuts away at the very fiber of our institutions and all that we as a nation have held dear for 245 years. The cancer is here and must be removed.

So, the next time you hear someone say that Black Lives Matter, stand firm and reply, "Yes, they do, because all lives matter, but liberty matters more!"

LIBERTY MATTERS MORE!

★ ★ ★

A Sailor's Reflections on Race and the Navy

By Brent Ramsey

I HAVE PREVIOUSLY EXPLAINED THE CONCEPT OF "SHIPMATE" through the recounting of the Navy career of my father.[235] He joined right after Pearl Harbor and served 27 years, retiring as a Master Chief in 1969, shortly after I was commissioned.

In 1948, President Harry Truman ended segregation in the military. Growing up in the late '40s and early '50s closely observing Dad's behavior and attitudes, I note he was an obedient Sailor and accepted everyone as they came, without regard to race. In 1955–1957, he was "pushing boots," and he treated all his recruits exactly the same, regardless of race or where they came from. The closest thing I ever heard him say that even hinted of a racial stereotype was that Black recruits did not know how to swim. It was not a criticism—just an observation that puzzled him, because, as a recruit-company commander, his job was to turn civilians into Sailors, and *everyone* had to be able to swim!

Ours was a Christian home, where one of my favorite Sunday School songs from my earliest memories is "Jesus Loves the Little Children," where the verse goes, *"Jesus loves the little children, all the children of the world, red and yellow, black or white, they are precious in his sight, Jesus loves the little children of the world."* The culmination of the civil rights movement in 1964 with the passage of the Civil Rights Act was celebrated in my household. And it was already the standard that the Navy had long since practiced. Justice was served!

In the late '60s, the biggest political issue at my university was anti-Vietnam war protestors, with Students for a Democratic Society (SDS) protesting us military types being allowed at university. After commissioning and school, I reported to the *USS Richard L. Page* (DEG-5) in 1970. *Page* was named after a Naval officer who in the Civil War became a Confederate brigadier general. No one, including black crew members, cared the least bit about the ship being named for a Civil War officer. Apparently hidden from my view was the torture and anguish our Black shipmates endured over the name of the ship (sarcasm alert!). The modern phenomenon of Blacks' lives being ruined at the very thought of the Civil War is a new invention of the fevered minds of academia that did not exist when I served.

The very best Sailor on *Page* worked for me. He was a first-class petty officer and the ship's Master-at-Arms as well. He happened to be Black. He set the example of squared-away Sailor for the whole ship, and nobody gave him any guff about his skin color. He was liked and respected by all. He was also a black belt in karate. The subject of White supremacy or systemic racism did not exist. Everyone was trained to a unity of purpose, which was to serve the ship and our shipmates and to make sure we were ready to fight and win in our nation's defense. Our entire focus was on the Soviet Union and the palpable, daily threat that the Soviets represented to America even in the western hemisphere. The crew was laser focused on the mission of being proficient at finding and neutralizing Soviet subs . . . and we were good at it, as the Navy Unit Commendation that I earned attests. Race was irrelevant to executing our mission and simply did not come up.

From Midshipman to CAPT, from GS-07 to GM-15, at 14 duty stations, and in a dozen locations, I found the conditions the same as on *Page*, Sailors and commands whose focus was on the mission and whose minority members shared the same values and qualities that everyone else had. And, even very early on, at my second duty station, I worked closely with an Air Force colonel who happened to be Black and thought absolutely nothing about it. Every duty station had Minority members, including senior officers. The numbers may not have been large back then, but the Navy was already diverse in racial makeup. I don't recall a single instance of racial tension or strife anywhere I was stationed in almost 40 years' combined military and civilian service. During Vietnam, rare instances of racial strife occurred elsewhere, but these were isolated instances. Race was not a significant factor in Navy culture or readiness. I never saw any discrimination or even heard of complaints of it in my entire career, much of which was in Mississippi. And, from my first tour on, all over the nation, every command had Minority members as part of the crew. Not only did I not experience any of the above, but also no one I know reported anything different in their own experience. The Navy I served in was focused on mission above all else. The Navy was not focused on solving the nation's residual race-related or cultural problems, nor should it have been. The Navy was a meritocracy by necessity, as our enemy was serious, implacable, and highly skilled at the art of warfare, and we had to be the same or better to prevail in the event of conflict.

I am not saying that the Navy did not have race-related problems. No doubt there were racists in the ranks and in some cases, discrimination occurred that was both illegal and harmful to the mission. But the institution of the Navy was committed to equal rights, as was the law, and, largely, discrimination was and is rare in the Navy, and those who discriminated were not tolerated. Up until recent years, no one had ever heard of Critical Theory or Critical Legal Theory or Critical Race Theory or Diversity, Equity, and Inclusion or Ibram X. Kendi's nonsense that you have to practice discrimination to make up for past discrimination.

When did it become an article of faith in our nation and in our military that the exact demographic makeup of the nation had to be matched in every other institution, including the military? What science supports that goal? If that were legitimate, why isn't everyone all up in arms that the NBA is 72% Black when the national demographic is 13% black? This is accepted as perfectly normal, but is the Navy officer corps having only 8% Black officers versus the 13% Black demographic nationally really a problem?

The Navy is an Equal Opportunity Employer and has been for a very long time. Are we absolutely free of race-related problems? Probably not. Idiots and bigots, despite our best efforts, do, at times, still join and may cause problems. When those surface, those people should be punished and discharged. According to a senior attorney with DoD known to this author, as recently as 2021, survey results from all of DoD (military and civilian) conducted by the Defense Human Resources Office, Office of People Analytics, overall, only *2 percent of DoD personnel are concerned about hate crimes or racism.* This is direct evidence that racism is not the problem being portrayed by military leadership or those with a political axe to grind promoting progressive ideology. The Navy should not be an experimental proving ground for politics. Those who join the Navy should be admitted based on qualifications, merit, and motivation. The color of one's skin should have nothing to do with it.

It is not the Navy's mission or responsibility to solve residual cultural problems with some Minority groups that may still exist. Those problems are America's to wrestle with and solve. Large majorities in the nation do not consider race to be a major problem, nor do they favor race preference for college admissions, including to the military academies. According to Pew Research, most Americans favor interracial marriage, and the percentage of such marriages continues to rise each year. In a few decades, most of America will be of mixed race. To ask the Navy to solve residual racial problems that the nation has apparently not been able to fully solve is ill considered and a distraction from the mission of having ready Naval personnel and forces in order to fight and win our nation's wars. There is no place in the Navy for social engineering. It will lead to defeat in battle.

This article originally appeared at AmericanThinker.com[236]

★ ★ ★

A Sailor's Reflections on Race and the Navy, American Thinker, 21 January 2023

The Navy's Misplaced Priorities Versus Core Navy Priorities

By Brent Ramsey

A CONFLICT WITH CHINA IS LIKELY IN THE NEAR TERM, and, fundamentally, that conflict will be maritime in nature. China's geography dictates the conflict will be at sea. Chinese early objectives are conquest of Taiwan, Japan, and the Philippines. Thus, China's provocative actions are manifest against those nations. Non-kinetic skirmishes are already so commonplace that they barely attract attention in the West. Tensions are vastly different on the other side of the world. Asian nations are buying billions in advanced U.S. weaponry to get ready for the inevitable. The Middle Kingdom views its manifest destiny as to rule the world, and that starts with conquering its nearest neighbors.[237]

Navy leadership should be laser focused on preparing for the coming conflict. They are not. In recent years, from Obama's Presidency to the present, the Navy has gone woke in a big way and is rudderless in preparing for the coming challenges. First let us examine the new woke Navy. Then, we will examine Navy basics like ships, leadership, and readiness.

The new Woke Navy:

- The former CNO recommended all hands read Ibram X. Kendi's book *How to Be an Anti-Racist*, a book openly advocating racism against Whites. Not explained is how reading this book makes for better Sailors. When confronted by Congress members about his choice of this book due to its divisive and political nature, he refused to remove it from his reading list. This is pure politics and has no place in our Navy.

- The next CNO will likely be a female officer with a degree in Journalism. So much for that STEM requirement, as apparently that is not important in senior officers. Selection of the Navy's leader based on gender is dangerous to our future success at sea in war, but nary a word is heard from any serving officers on this "political" appointment. Where are the senior Admirals who will call out partisan politics dictating to the Navy? Does anyone remember the revolt of the Admirals, where principle meant more than career?[238]

- The Navy Inspector General recently confirmed that the former Superintendent at the Naval Academy made false statements about a midshipman he wanted to expel. The midshipman's offense ... he made private statements in support of the police, specifically his parents, both police officers, during the protests and riots in the aftermath of the death of George Floyd. Fortunately, justice prevailed, and the midshipman was commissioned and is now a serving officer. Yet the man responsible for enforcing the USNA Honor Code is left to slide on the same honor code it is his manifest duty to enforce.

- The Navy is in a crisis to get people to join the Navy, falling short of the numbers needed by thousands. Solution: Let's use an active-duty sailor who performs as a drag queen to jack up recruiting. Elaine Donnelly of the Center for Military Readiness captured it perfectly ... "The U.S. Navy Steps on a Rake with Drag Influencers."[239]

- Maury Hall at the U.S. Naval Academy is renamed. So, also, was the former USNS *Maury*, an oceanographic ship. Matthew Fontaine Maury, for more than a hundred years, was considered the father of the science of oceanography and was also a "professor of physics at VMI and had an internationally acclaimed career as an oceanographer, astronomer, historian, meteorologist, cartographer, geologist, writer, and commander in the United States Navy. Accordingly, he is affectionately remembered as the Pathfinder of the Seas, Father of Modern Oceanography and Naval Meteorology, and Scientist of the Seas."[240] Not good enough for the politically correct, so it was necessary to erase him from Navy history. How many of America's

Black citizens ever even heard of Matthew Fontaine Maury? Exactly who benefitted by eradication of his name and reputation?

- In recent remarks, the Secretary of the Navy announced that dealing with climate change is a top strategic priority.[241] He said fighting climate changes is an "all hands on deck" priority. China, not so much.

- Stemming the scourge of sexual assault has been a high priority for a long time. The result, higher rates of sexual assault than ever, with the Navy leading the way with the highest rates. Could it be accentuating sexual topics with gay pride celebrations, embracing so-called transgender people, including paying medical costs to transition is stupid . . . essentially promoting more overt, more aggressive sexual behavior? Instead of service over self, the Navy promotes identity over service.

- The plague of suicides continues apace in the Navy with no solution in sight despite CNO after CNO promising to make suicide prevention a high priority. During Covid, three times as many Sailors killed themselves than died of Covid. In 2021, there were 328 military suicides. During the 18-month period from March 2020 to August 2021, there were only 43 deaths from Covid.[242] Tell me why the Navy mandated all hands to get the vax—despite the extremely low risk—and ignored the suicide epidemic. Had the thousands of hours spent persecuting Sailors with religious objectives to the vax been devoted to helping despairing Sailors, hundreds of deaths could have been prevented.

The essential Navy: Ships, readiness, and leadership:

- Heritage Foundation, in its 2023 Index of Military Readiness, finds the Navy to be "Weak," down from "Marginal" in its last report.[243]

- A deluge of senior-leadership failures plagues the new woke Navy. A week does not go by without more skippers being relieved for cause. There is obviously something wrong going on here—exactly what is not clear. During a period of only three months in 2022, the Navy fired 13 commanders and refused to comment on the causes.[244] In his confirmation hearings, former Secretary of the Navy Braithwaite said, "It saddens me to say: the Department of the Navy is in troubled waters due to many factors, primarily the failings of leadership."[245] Coming from a retired Admiral, this should have been a wake-up call. Sadly, he served only a few months until the election brought us a new Secretary, an identity-politics pick, Secretary Del Toro, a retired Commander, of Hispanic ethnicity.

- Everyone remembers the tragic collisions of the *USS Fitzgerald* and *USS McCain* in the Pacific in 2017. It was a wake-up call, we were told. The surface Navy was in decline, and Sailors were dying because of it. Navy leaders pledged to fix it. It didn't happen. The latest INSURV report shows the deplorable readiness of our ships. This is on Navy leadership and nothing else.

SURFACE							
Functional Areas (Ships Inspected)	2017 (17)	2018 (18)	2019 (36)	2020 (25)	2021 (17)	2022 (33)	2022 Comparison to 6-Year Avg
Main Propulsion	0.81	0.79	0.67	0.73	0.78	0.76	ABOVE
Anti-Sub Warfare	0.96	0.85	0.88	0.81	0.87	0.86	BELOW
Communications	0.82	0.79	0.75	0.77	0.75	0.74	BELOW
Information Systems	0.60	0.63	0.55	0.56	0.56	0.59	ABOVE
Aegis Weapon Systems	0.77	0.70	0.66	0.62	0.69	0.69	ABOVE
Mine Warfare	0.91	0.78	0.76	0.73	NA	0.78	NEUTRAL
Operations	0.86	0.82	0.78	0.77	0.71	0.74	BELOW
Weapons Systems	0.80	0.73	0.69	0.67	0.71	0.63	BELOW
Auxiliaries	0.84	0.84	0.82	0.83	0.80	0.78	BELOW
Electrical	0.73	0.69	0.70	0.76	0.74	0.80	ABOVE
Damage Control	0.80	0.79	0.79	0.77	0.77	0.79	NEUTRAL
Deck	0.83	0.78	0.76	0.69	0.75	0.78	ABOVE
Navigation	0.87	0.87	0.86	0.85	0.82	0.84	BELOW
Aviation	0.63	0.75	0.49	0.48	0.61	0.45	BELOW
Preservation	0.84	0.82	0.84	0.84	0.81	0.78	BELOW
Ventilation	0.76	0.75	0.77	0.78	0.61	0.69	BELOW
Environmental Protection	0.85	0.81	0.76	0.76	0.73	0.65	BELOW
Medical	0.94	0.93	0.91	0.92	0.93	0.95	ABOVE
Supply	0.76	0.78	0.80	0.80	0.83	0.80	NEUTRAL
Habitability	0.80	0.78	0.78	0.78	0.74	0.80	ABOVE
NAVOSH	0.78	0.70	0.72	0.72	0.75	0.72	BELOW

Figure 5.1 6-Year Surface Functional Area Scores

- In 2020 the *USS Bonhomme Richard* burned pier side. It was a total loss. A capital ship with a replacement cost of $3B was lost due to institutional failure to perform a fundamental Navy function since navies have existed in the history of man . . . effectively fighting a fire on a ship. The result? Essentially nothing—a JAG manual investigation that never conclusively proved the origin of the fire and the ends of the careers of a few of those involved. Nothing of substance was the result, and the Navy was out a $3B ship. A full summary of the failures of leadership in this debacle are contained in a report by *Defense News*.[246]

- The Navy has finally given up on the LCS class. Billions were wasted on a ship that never reached mission capability, and now ships are being decommissioned after as little as 8 years in commission. The expected life span of an LCS was 30 years. The Navy is still building LCSs even though it is decommissioning earlier-built ships after a few years in commission. No wonder Sailors called it the "Little Crappy Ship."

- Zumwalt class is still searching for a mission. Billions for a 3-ship class that the Navy still does not know what to do with. The latest in a series of ideas on what to do with the ship is to arm it with hypersonic missiles—which the Navy does not yet have. The ship has been retrofitted with VLS missiles but packs less punch than the Burke Class, which is half its size (80 missiles versus 96 on the Burke class).

- The Navy has failed to produce a Congressionally mandated Shipbuilding plan for years. The result is that Congress has decided to create its own Commission to produce a plan for the Navy.[247]

- For decades, the Navy was required to maintain 38 amphibious ships to support the mission of the USMC. Now the USMC wants 31, and the Navy and Congress disagree, planning to take that number even lower.[248] Confusion reigns, it seems, in even knowing the requirement for the number of ships in light of the Commandant's remaking of the USMC. Bing West, probably the most decorated and experienced living Marine, had this to say: "General Berger concocted his concept in secret, not consulting the retired four-star community that, appalled by his extensive cuts, has united in opposition. Multiple articles have been written laying out the irremediable defects in the anti-ship strategy. To sink some Chinese warships in a dubious war scenario, Marine resources and organizational cohesion have been severely damaged. General Berger's injudicious change of direction will adversely affect Marine warfighting capabilities, internal morale, and recruiting for years to come."[249]

- The Columbia-class submarine is behind schedule.[250] The GAO has issued reports that outline the risk inherent in the program to replace our aging Fleet Ballistic Missile submarines, the oldest of which is already 41 years old. To achieve the milestones that it has met, it has robbed personnel from another critical submarine program, the Virginia-class advanced SSN.

- The USS Ford took 9 years to build, cost $13B, and still has operational-capability problems with both launching and recovering aircraft 6 years after being put into commission.[251]

Conclusion

Whether it is manning, managing the personnel, effectively leading and taking care of Sailors, or actually building and operating ships needed to fight our enemies, the Navy is doing a poor job. The record is clear. The foregoing is common knowledge among thousands of retired Naval personnel. There is a growing chorus of voices from retirees and organizations like Calvert Task Group, STARRS, Center for Military Readiness, USNA-at-Large, Restore Liberty, Veterans for Fairness and Merit to alert the public. Heritage Foundation, Judicial Watch, Armed Forces Press, Legal Insurrection, and *The Federalist* have also sounded the alarm as to the danger if we don't change course. It is past time for citizens to wake up to the danger of an ineffective Navy and call upon the Congress to mandate changes needed to restore the Navy to high-quality leadership, effectiveness in shipbuilding, and readiness. Equally important, get woke politics out of the Navy! *This article appeared in Real Clear Defense.*

★ ★ ★

The Navy's Misplaced Priorities Versus Core Navy Priorities, Real Clear Defense, 6 June 2023

The Great Diversity Hoax

By Jim Tulley

"Why" is a wonderful, troublesome, puzzling, and sometimes irritating word. When asked by children, it often elicits a response of, "Because I said so." In other situations, it might cause us to think. So, when someone says, "Our diversity is our strength," why does no one ask "Why?"

If the statement means diversity of thought, who could disagree? This country has practiced such for more than 200 years. It is central to the structure of our federal government, as the Legislative, Executive, and Judicial branches function each in their own unique ways. Furthermore, all 50 states govern in different ways that work best for their unique circumstances. If it's diversity of culture, there is no argument: America is one of the most diverse nations in the world. In an open society, it is almost a given.

However, racial diversity for its own sake weakens the effectiveness of any organization because skin color is a superficial characteristic of a human. It's clear that the United States Naval Academy is purposely setting out to make the Brigade of Midshipmen "look like" the Fleet or perhaps the nation. To promote this goal is a Naval Academy Office of Diversity, Equity, and Inclusion (ODEI). Does the Superintendent believe that diversity means "looking like" America?

The Navy is not the only service with this approach. Social experimentation in lieu of merit is now prevalent throughout the U.S. military. In May of 2023, the *Daily Caller* reported, "The U.S. Air Force abandoned an experiment aimed at boosting pilot-training graduation rates for women and minority pilots after the 2021 initiative failed to achieve the intended results."[252] How could it? The explanation is quite simple; anytime another metric is introduced into any selection process that trumps merit, merit will suffer. Unless all candidates are of equal capability, the net result must be a reduction in merit-based qualifiers that had otherwise been used. There is no other possibility. So, when it is claimed that "Diversity is our strength," what is the strength we are attempting to achieve via diversity?

The United States Naval Academy's Office of Diversity, Equity, and Inclusion (ODEI) Mission web page[253] states goals of making the Academy an "inclusive campus," ensuring "equitable access" and addressing the "challenges of underrepresented populations." This reads like political-correctness dogma at an institution where the real mission is to train combat leaders. So, Admiral, please enlighten us. How and why does your new, diverse culture make our Navy a stronger fighting force? Where is the evidence?

Some say that Sailors and Marines are best led by officers of their own race. Why is this true? If that were so, why don't we have White-crewed ships, Black-crewed aviation squadrons, and Latino-crewed submarines? In 1948, President Truman signed Executive Order 9981[254] that ended segregation in the American military. And, in 1954, the U.S. Supreme Court, in *Brown Vs. the Board of Education,*[255] decided against the idea of separate-but-equal in our classrooms. Though aimed at public education, surely that concept applies just as much to our military. We are and should be one Naval Academy, one Navy and one Military. To think otherwise only serves to perpetuate the hoax.

In his book, *The Dying Citizen,*[256] Victor Davis Hanson explains that the duty of what he calls campus "diversity commissars" is to "monitor the precise racial makeup of the labor force, to adjudicate the authenticity of individual racial pedigrees, to reeducate the majority of the White population about its toxic insidious privilege, and to ensure that curricula and communications include proper vocabulary and phraseology." Since the Naval Academy's ODEI functions outside the regular chain of command, why has the Academy created what amounts to an Office of Diversity Commissars when the Naval Academy is already diverse by any standard except ODEI staff?[257]

The professional sports world does not bow to diversity. In 2020, the *New York Times* reported[258] that the NBA, WNBA, and NFL were composed of 83%, 83%, and 74% "players of color," respectively. Is that diversity? Imagine the response from league Commissioners to, "Your league is not sufficiently diverse. You need to look more like America!" Can you hear the laughter? Professional sports are about winning. They care nothing about skin color, only about drafting and keeping the best players. Should our military function any differently? If so, why?

All of the four major Service Academies have offices of Diversity, Equity, and Inclusion. The Air Force Academy even offers cadets an optional Diversity and Inclusion minor.[259] So why do these prestigious military institutions now seem more concerned with ensuring "equitable access'" and addressing the "challenges of underrepresented populations" than they do military readiness and unit cohesiveness? These are cultural problems in our society that should be addressed elsewhere. "Equity" means "equality of outcome," usually carried out via explicit or implicit quotas. Quotas negate merit and result in overlooking some of the most-qualified candidates. It is contrary to the essence of Dr. Martin Luther King's message of judging all people by the content of their character, not the color of their skin.

There is another factor that is closely related to the all-too-obvious military-diversity factor. There are currently 84 Historically Black Colleges and Universities (HBCU) in the United States, offering four-year undergraduate degree programs and having a collective enrollment exceeding 250,000 students.[260] Of these institutions, only eight have anything like an Office of Diversity, Equity, and Inclusion, and most of those acknowledge only one or two of the three elements. One of the eight has a Center for Promoting Social Justice, and another has a Center for DI and Social Equity. Another, Huston-Tillotson University, brags on its web site, "HT is one of the most diverse institutions of higher education in Texas, with a population of 70% African American, 19% Hispanic, .05% Anglo, .04% International, and the balance Asian and Native American, according to 2014 data." If diversity is such a priority, why is it largely non-existent at the HBCUs?

Why would these HBCUs not aggressively seek the kind of diversity that is sought at other prestigious colleges and universities? In addition to the Service Academies, Harvard and Princeton both have offices of Diversity, Equity, and Inclusion, as does Yale University and Duke University Medical Schools. Stanford University has a program called "Diversity Works," and MIT has a Diversity Dashboard. All of these offices and programs are proudly displayed on the university websites. Apparently the nation's HBCUs have not embraced the concept. Looking at their websites, one could easily conclude that these schools are century-old models of racial segregation. Why? Does "diversity" now mean something other than what we have been led to believe? Perhaps diversity is a one-way street and a throwback to Affirmative Action run amok.

One final comment by Dr. Hanson from *The Dying Citizen* is pertinent.

> "Once Americans embrace such ethnic chauvinism and identify by superficial appearances . . . even on the pretext of correcting past wrongs—then embarrassing contradictions, ironies, contortions, and paradoxes are inevitable . . . The result is that an entire generation of youth has grown up and been educated on the now-mainstream premise that their ethnic and/or gender identifications define who they are at the expense of their commonality as Americans."[261]

So, to all leaders who claim it to be so, we ask again, why is racial diversity a strength? What about diversity makes our Navy and our military a stronger fighting force, better prepared to win our nation's wars? In truth, racial diversity for its own sake serves only to dilute merit and weaken our effectiveness. It is a fact at the Academies and in other universities that, in order to achieve their diversity goals, students with lower academic achievement by multiple measures (test scores, GPA, etc.) are admitted, lowering the overall academic quality of the student body to achieve an arbitrary skin-color goal. It is a misguided hoax perpetrated by those who are determined to sacrifice merit on the altar of political correctness. The United States Service Academies must be dedicated to strengthening our commonality as Americans defending our Constitution against all enemies, fighting with courage and resolve, and, most important, winning. An unbalanced focus on racial diversity is a distractive, cancerous hoax and must be excised from our thinking.

★ ★ ★

The Navy and Diversity, Equity, and Inclusion

By Brent Ramsey

DEI ADDS NOTHING OF VALUE TO OUR NAVY. It just wastes resources and causes division within the ranks.

The Navy is all in for DEI. At the Navy website, if one clicks on "Who We Are," one of the first things that shows up is "diversity and equity." It must be important to show up so prominently.

The Navy Diversity and Equity website page says: "I AM A SAILOR. WE ARE A TEAM. THIS IS OUR NAVY."

> "When Sailors feel included, respected, and empowered, they will be more ready to win wars, deter aggression, and maintain freedom of the seas."
> —ADM Mike Gilday, Chief of Naval Operations.

The Navy offers no evidence to support the CNO's statement. It is not clear how the statement actually relates to diversity or equity. It does mention feeling "included"—a form of "inclusion"—but in a vague sort of way. In my 34 years in uniform, I don't recall at any time being asked my feelings. Must be a new thing. I doubt the People's Republic of China military leaders ask how their people are feeling. The Navy I served in stressed toughness, stamina, perseverance, physical fitness, strength, honor, courage, and commitment. My feelings were secondary, and that was well-understood by me and my shipmates. I have been literally eyeball to eyeball with the Soviets in the North Atlantic tracking a Yankee-class submarine. Rest assured that former enemy had our full attention, and our crew devoted no time to feelings. I have no doubt our current adversaries are just as potentially dangerous to our way of life as was the former Soviet Union. The Navy ought to focus on our real threats instead of touchy-feely nonsense like Sailors' feelings. The Navy is a combat force whose job it is to break things and kill people when and where called upon to do so. Those whose personality or psyche demand constant attention to "feelings" probably ought to find something else to do.

Let's look at diversity

How diverse should the Navy be? It doesn't say. The fact is, the Navy is already about as diverse as any institution anywhere. The 2021 DOD report on demographics for Navy shows (percent):

	Native-1[262]	Asian	Black	Native-2[263]	Multiracial	Unknown	White
Enlisted	2.0	6.0	19.4	1.3	6.7	4.4	60.3
Officer	1.1	5.9	7.9	0.5	5.1	3.3	76.2
USA (2020)[264]	1.1	6.0	12.4	0.2	8.4	—	

DOD's 2021 demographics shows the Navy has 114,100 enlisted personnel and 13,361 officers who identify as a Minority. The Navy is already diverse: 37% of the Navy is from a measured racial demographic group, leaving 63% who identify as White, which matches the percentage reported nationally in the 2020 Census. That is more diversity than the national average. An anomaly hidden from view is the Hispanic segment of the military. You will note that there is no category for Hispanic in the table above. Actually, 17.7% of the active DOD force is Hispanic, but you would never know that from the figures above, because it is not reported. The U.S. census does not count Hispanics as a minority. Instead, they are considered a different ethnic group. According to Pew Research Center data published in 2021, 58% of Hispanics consider themselves to be White with the remainder identifying as some other color.[265] This unique accounting obfuscates the fact that the U.S. military is even more diverse than appears to the naked eye when viewing DOD's reports of racial groups. In fact, more than 350,000 Hispanics serve in the military and add considerably to the diversity of the total force.

Other than a slight underrepresentation of Black officers, the Navy is at or exceeds the national demographic of racial diversity. This begs the question, what is all the fuss about diversity? The Navy is diverse! There is no need to become even more diverse than the Navy already is. Millions of dollars and precious time are being devoted to a problem that is outside of the Navy's ability to control. Nay, you say. . . . what about the shortfall in Black officers? What about it? The Navy is devoting a lot of time and effort into trying to recruit Blacks to join the Navy to become officers, but the gap remains. One must ask the question, "Why?" It is not because of discrimination, because there are already thousands of Black officers, so there is obviously no barrier to Blacks becoming officers. What is lacking is Blacks who want to become Naval officers. No amount of incentives or handwringing over a slight underrepresentation of Blacks in the Naval-officer ranks is going to change a situation that obviously has other causes. Even the small shortfall is actually not the Navy's fault, as is brilliantly and eloquently analyzed and explained in detail by CDR Phil Keuhlen in Task Force One Navy Final Report: "The Emperor's New Clothes Redux."[266] According to government graduation data, there are more than 200,000 Black college graduates each year. The problem is that very few of these qualified people have an interest in serving in the Navy. We ought to be curious about why. It is not because of discrimination, as DOD's own internal reports document that fewer than 2% of the 3.4 million who serve in DOD consider racism to be a problem.

Let's look at equity
The Navy DEI webpage goes on to say "Putting on a uniform doesn't mean sacrificing who you are. America's Navy values diversity, equality, and inclusivity—striving to build a community of service members who accurately reflect the rich makeup of our country. Our belief is that, with hard work and determination, anyone, from anywhere, has the power to be successful in the Navy." The Navy uses the word "equity" earlier twice, and, then, in the next paragraph, the word "equality" is used. Which is it? These two things are not the same. "Equity" means equal outcomes regardless of merit. "Equity" is a term associated with social-justice advocates who want equal outcomes for everyone regardless of merit. Is the Navy really advocating equal outcomes? How does that even work in a military organization with

rigid technical requirements that dictate practices for safety reasons and for warfighting effectiveness? Military organizations are rigid, structured, top-down organizations whose fixed chains of command and methods demand uniformity, strict discipline, and consistency in order to function with any kind of efficiency. Adopting equity in a military chain of command is deadly and dangerous—and will get people killed. "Equality" is equal opportunity based on merit, which is and has been the law of the land for a long time. "Equality" should be the Navy's mantra, not "Equity." The Navy's use of both terms may confuse those who visit their website. Is the Navy using both terms to conflate the two words into meaning the same thing in order to stimulate interest in the Navy?

Let's look at inclusion
What does that even mean? In one section "putting on a uniform" is the first thing said. The word "uniform" is telling! It absolutely proves the Navy strives for uniformity . . . that's why you wear a uniform. Everything top to bottom in the Navy is about uniformity. Ships and aircraft are built to uniform specifications. The way things are done is based on time-tested, uniform best practices. Individuality is not allowed because that leads to bad outcomes and people dying. Putting on the uniform is absolutely about giving up your individuality while that uniform is on. Most of the time, you absolutely must sacrifice who you are to join the Navy. On duty, you must conform to the Navy's uniform standards for dress, behavior, operational excellence, and a thousand other things. When you are off duty, then you have time to be yourself. While on duty, you belong to the Navy—you are included by wearing a common uniform, the same one at the same time as everyone else does. When you get up in the morning, you look at the Plan of the Day, and it tells you what the uniform of the day is. How inclusive is that? We are all exactly on the same page, and the uniform reminds you every day that you are part of something bigger, something important. What does the use of the word "inclusion" imply? Are there some fields in the Navy that are open exclusively to white people? Of course not. The Navy's use of the word is purely political, and politics should have no place in the Navy. It is used to show that the Navy has gotten with the program and is using the same language as the larger society and as desired or even dictated

by the political left. But how does using politically charged terminology make for a better Navy? It doesn't! It has no place in the Navy and serves only to introduce doubt and dissension, and make the Navy a less-effective force. A cursory search reveals the popular origin of the use of the word in today's parlance. Where you find the term used is in colleges and universities, where the political philosophies of the left are preached, including in Critical Theory, Critical Legal Theory, and Critical Race Theory.

Conclusion

There is no place in the Navy for the politics of DEI. The Navy is already diverse. The data is not in doubt. Equity has no place in the Navy. The Navy's moral and practical foundation is a merit-based organization focused on being able to fight and win the nation's wars at sea and projecting power ashore. Equity undermines merit. Dilution of merit in favor of equity will get people killed. Anyone telling you anything different is mistaken and should be ignored. Inclusion is a phantom and is wordplay from the political left. When you join the Navy, you are included by the very uniform you wear, which is the same for everyone. The UCMJ guides your conduct and demands fair and equal treatment of all. The oath you swear to uphold has as its foundation the Constitution. Equal protection is guaranteed in the 14th Amendment and has been upheld by the Supreme Court many times, including recently in the Harvard and UNC cases on college admissions. Nothing more is required. DEI adds nothing of value to our Navy. It just wastes resources and causes division within the ranks. *This article appeared in Real Clear Defense.*

★ ★ ★

The Navy and Diversity, Equity, and Inclusion, Real Clear Defense, 18 July 2023

Wokeness Is Antithetical to Military Service

By Tom Klocek

"Wokeness" is all about oneself. It is political correctness on steroids. It rewards conformity to ideological perspectives rather than conformity to real performance. It does not look at others, their needs (except when "virtue signaling" or trying to make yourself appear to be taking the moral high ground), or how to work with them. The basic principles of being "woke" include the lack of personal responsibility for one's actions, the need for safe spaces when there is a hint of adversity, and the right to be offended at the slightest trigger.

This ideology has infected America's military and is wreaking havoc among our troops. Young adults entering the Armed Forces are indoctrinated into being afraid to think for themselves. Self-sacrifice is foreign to them. They are being pushed to believe that their nation, their upbringing, their families, and their history are all based on hate. These characteristics are directly at odds with service to one's country.

Wokeness is antithetical to everything the military represents. It is divisive and isolates one group from another based on immutable characteristics, including ethnicity and sexual orientation. You cannot build an effective team if its members are in conflict with each other. Teamwork requires unity, not divisiveness, which is prejudicial to good order and discipline.

Moreover, the military services have led the way in fighting racism and fostering equality of opportunity. For decades, the military has provided equal pay for equal service. Now, however, the military Academies have affinity groups that are inherently divisive, grouping people by external characteristics, such as skin color.

Military service is built on performance. The mediocre pilot is not going to fare well against an enemy who has honed his flight and dogfight skills against difficult opponents. We don't want our fighters to be passable—we want them to be better than the enemy.

Another major characteristic of military service is self-sacrifice. Military members give up time with their families to deploy for months at a time, even more in crisis situations. They put their lives on the line. One doesn't have to be in a combat zone to be in dangerous situations. For example, just going to sea is an inherently hazardous situation.

Consider some of the mottos and sayings that characterize the sea services, some of which we memorized during our Plebe Year at the Naval Academy. "I have not yet begun to fight" (John Paul Jones, 1779). "Don't give up the ship" (CAPT James Lawrence, War of 1812). And here's one that is not well known but which is exemplary of warfighters: "Always pray, not that I shall come back, but that I will have the courage to do my duty" (Lt. Anthony Turtora, Marine Corps, Guadalcanal).

The sea is not a place for wimps. You don't get a participation trophy: you survive, or you don't. And to survive, you need courage, strength, teamwork (unity), and perseverance. In an emergency, an almost-daily occurrence for ships at sea, teamwork is the difference between life and death. Any distraction due to a trivial woke sensibility can be deadly. Just ask the crews of the *USS Fitzgerald* and *USS McCain*, who continue to mourn lost shipmates.

The men who signed our Declaration of Independence were not looking for a "safe space." The Constitution guarantees to protect every state against invasion and empowers Congress to "raise and support armies" and to "provide and maintain a navy." Our Founding Fathers knew the hostility of the world at large and thus provided for the protection of the nation. The woke agenda weakens our defense by changing the priorities of the services from protecting the country to catering to individual members within the services themselves, thus diverting attention from issues of readiness and leadership.

Most people who join the military do so because they believe there is something greater than themselves. Even during the draft, the underlying concept was service to the nation as a necessary aspect of citizenship. Individualism had to take a back seat. The essence of wokeness, on the other hand, is focused on the individual—the exact opposite of what our military needs to function effectively.

Consider this quote from Charles Michael Province, U.S. Army:[267]

> *"It is the soldier, not the minister, who has given us freedom of religion.*
> *It is the soldier, not the reporter, who has given us freedom of the press.*
> *It is the soldier, not the poet, who has given us freedom of speech.*
> *It is the soldier, not the campus organizer, who has given us the freedom to demonstrate.*
> *It is the soldier, who salutes the flag,*
> *who serves beneath the flag,*
> *and whose coffin is draped by the flag,*
> *who allows the protester to burn the flag."*

Can today's military continue to claim that honor?

★ ★ ★

Danger Close: People's Republic of China

By Brent Ramsey

We are at war, but most Americans are blissfully unaware. Folks are busy living their lives, attending school, working, raising kids, pursuing dreams, or if retired, leisure activities, visiting grandkids, enjoying life. Most are ignorant of the malign designs of the Chinese Communist Party (CCP) of the People's Republic of China (PRC).

China is ruled by CCP Chairman and President Xi Jinping. His word is law. His goals are anathema to freedom-loving people everywhere. The people of China are helpless pawns. They have no rights. Government may lock people in their dwellings for months on end, with no reason and no recourse, as they recently did. There is no freedom, no autonomy. Everything, everyone in China is subordinate to the will and whims of the CCP and its absolute ruler, President Xi. I have written extensively on the threat of the CCP.[268] China expert Gordon Chang[269] of Gatestone Institute warns that China is an extreme danger to Americans. Dr. Michael Pillsbury of Heritage Foundation predicts China will overtake America as the world's superpower.[270] Former Trump National Security Advisor Lt Gen H. R. McMaster has sounded alarms about the PRC in Congressional testimony.[271] Dr. Pillsbury concludes there is a 70% chance that the PRC will dominate the world. Unfortunately, too little heed is paid to these warnings. We don't want a world ruled by China.

Decades ago, Congress established the US-China Economic and Security Review Commission. That non-partisan body has been doing yeomen's work warning the nation of the CCP/PRC threat. Their sobering 2023 Executive Summary is well worth reading.[272] The full report lays out in stark detail the corrupt and evil ways the PRC is using to gain power, influence, and dominance in virtually everything. Experts consider China a peer or near-peer to the United States in both military and economic strength. A 2023 study by Australia Public Policy Institute documents China's leadership in 37 out of 44 fields.[273] Technology is crucial to economic and military strength. China is a leader in space, Artificial Intelligence, quantum computing, rare earths, manufacturing, college degrees, patents, pharmaceuticals, hypersonic missiles, shipbuilding, batteries, solar panels, manufacturing, port control, and many other fields. Senator Rubio issued a startling report, "Made in China 2025" on the CCP's plans to dominate in 10 critical sectors by 2025.[274] They are well on their way to this goal.

Many now predict a military confrontation with the PRC/CCP in the near term. It likely will be over Taiwan. I have written about that threat at CD Media.[275] President Biden and other national leaders state forthrightly that the U.S. would come to the aid of Taiwan. We have authorized billions for defense weaponry and promised more. TRANSCOM General Minihan recently warned that war will begin with China as soon as 2025.[276] What is the state of our readiness?

You might think our military will save the day. Heritage Foundation's Index of Military Readiness paints a grim picture. This year's report[277] warns in stark terms that our military capability and readiness has fallen. Heritage states, "In the aggregate, the United States' military posture can be rated only as "weak." Weak budgets, neglect of the military by both parties, and the politicization of the military has led to a shocking weakness—just when China is stronger than ever.

Our military is not ready to fight the PRC!
Both Heritage and the Security Review Commission are sounding the alarm! Exacerbating a bad situation, the politicization of our military, which started in 2009 under President Obama, is now on steroids! President Biden greatly accelerated what Obama began, mandating divisive, wasteful social-justice indoctrination. Critical Race Theory, Diversity, Equity, and Inclusion, gay-pride celebrations, drag-queen shows, women in combat units, indoctrination in what pronouns to use, and integration of transgenders are military-commander top priorities—it's consuming resources and

distracting from the core mission. With what result? Readiness is down. All services missed recruiting goals last year. Retention is down. Military-member suicide trends are unconscionable, and the problem is glossed over by leadership. Sexual-assault rates have skyrocketed 18% as shown by this report.[278] Our military is distracted and not ready. American youth are declining to join the military. Families who served generations are telling their children to avoid the military. The nation is waking up to this danger thanks to the efforts of veterans organizations who are documenting the problems. STARRS,[279] Calvert Group,[280] Restore Liberty,[281] Center for Military Readiness,[282] and Veterans for Fairness and Merit are sounding the clarion call to restore merit and traditional values in our military, including fidelity to the Constitution. Visit their websites and support their efforts. Members of Congress are warning of these dangers, too, such as Senator Rubio and Congressman Roy, who recently issued a scathing report outlining dozens of examples of the wokeness that has weakened and distracted our military.[283]

In 1941, at Pearl Harbor, the world saw the U.S. viciously attacked with our pants down. Almost 3,000 died on that single infamous day. In hindsight, it's clear that the U.S. ignored the warning signs that presaged that attack. Our failures in diplomacy and readiness led Japan to wrongly conclude that they could defeat us. That led to 111,000 American deaths and 250,000 wounded in the Pacific theater alone. Let's not repeat that type of failure with the PRC, with 10 times the resources and population of Japan plus nuclear weapons.

We must elect those who will strengthen our military. We must have a military that focuses on readiness alone. We must educate the American public on the dangers to America that the PRC poses. Congress must act to better protect us against the CCP's evil plans and strengthen our military. If our military is not ready to defend us, you might as well start learning Mandarin. In a stark warning, China's Minister of Defense openly discussed using biological weapons to kill Americans. Both the Department of Energy and the former head of the CDC confirmed that the Covid virus was developed in Wuhan, China. What else are they developing? Think of that. Every one of your loved ones could be dead in a few years in the coming conflict if we don't act now to preserve our military and its ability to protect and defend us. The time to act is now. *This article appeared in Armed Forces Press.*

★ ★ ★

Danger Close: People's Republic of China, Armed Forces Press, 19 March 2023

Connecting the Dots

By Jim Tulley

There is in common usage a phrase that most people know, "Where there's smoke, there's fire." In the past two years, we have seen much "Woke Smoke" surrounding a transformed culture at the United States Naval Academy, as well as at West Point and the Air Force Academy. Such observations have raised alarms among many retired senior-officer graduates of all three institutions. Much of this clear cultural transformation involves what can only be described as covert indoctrination of Midshipmen (and Cadets) in the tenets of Critical Race Theory (CRT). Is this smoke a sign of a deeper problem?

Some indication comes from the highest levels of government, including the U.S. Senate. Senator Tom Cotton (R-Ark.) told Defense Secretary Lloyd Austin on June 10 that "five hundred" military whistleblowers have reported being forced to receive "anti-American indoctrination" training, including critical race theory (CRT). This raises concerns about the military in general, but we are most concerned about the United States Naval Academy.

As we have become more aware of possible problems, a small group of graduates from the class of 1969 formed the Calvert Task Group to look more closely at the allegations, question the players, and take action as needed. We see the smoke and are concerned. The group is named in honor of Vice Admiral James Calvert, who was the Superintendent while the class of 1969 was at the Academy. He was our mentor, and his example of leadership has had a lasting effect on our lives.

In addition to the comment by Senator Cotton, we have seen a series of issues that might be called a set of "dots" that, when connected, paint a disturbing picture of the Naval Academy. These points of concern include but are not limited to:

- Inconsistent statements made by then-Naval Academy Superintendent Vice Admiral Sean Buck.

- Suggestions or Critical Race Theory (CRT) indoctrination present in a revised USNA culture, even noted in an outside organization, CriticalRace.org.

- The presence of a wide range of affinity groups based exclusively on race or culture.

- An Office of Diversity, Equity, and Inclusion (ODEI) that appears to function outside the historical Brigade Chain of Command.

- The illegal dismissal from the BOV of all the conservative appointees, as mandated by SECDEF, who did this without authority.

- Recent litigation that has also asserted negative comments made by the Superintendent

Taken on their own, any of these points would be cause for concern. Taken together, they paint a picture that cannot be ignored.

We are concerned about inconsistent comments by then-Naval Academy Superintendent Vice Admiral Buck. From an attendee at the recent reunions of the classes of 1971, 1976, and 2010, we are informed that Admiral Buck spoke about diversity and CRT. He denied that CRT is being taught at USNA and has had a team review past curriculum to verify. He went on to say that, in the past, there was a course with a different name that addressed the tenets of CRT.

The Superintendent was adamant that CRT was not being taught and went on to say that "it never will." Paraphrasing William Shakespeare, "Methinks thou dost protest too much." Admiral Buck's responses suggest that he recognizes CRT as a negative force in the Navy and at the Academy. But if he really believes that, why is it not being taught? That is to say, why are Midshipmen not being made aware of its Marxist history and the divisive concepts imbedded in its philosophy? Where there's smoke...

There are other suggestions from outside the Academy that CRT is a problem. CriticalRace.org is a resource for parents and students concerned about how Critical Race Theory and implementation of Critical Race Training affects education. It maintains a comprehensive database to empower parents and students to understand how CRT might affect students attending the recorded institutions. In its pages, it reports that, "Not all of the colleges and universities in this database have Critical Race Training." For those who do have such Critical Race Training, there are *varying degrees of such programming*, some mandatory, some not. For many schools, it's a continuum of programming, such as "*Diversity, Equity, and Inclusion*" and "implicit bias" training and programming, which does not easily fit into a Yes/No construct. We are concerned about the underlying meaning of "Diversity, Equity, and Inclusion" training at USNA. Where there's smoke...

In the report on USNA, CriticalRace.org states,

1. USNA has engaged in numerous actions in support of DEI efforts and anti-racist teaching. For example, USNA created the Naval Academy's Diversity and Inclusion Strategic Plan. USNA provides training on DEI throughout the year, in addition to countless resources. Annual training for all Midshipmen, leadership, faculty, and staff will be a requirement soon.

2. The Superintendent of the United States Naval Academy said, "Training sessions about the importance of diversity to our institution are scheduled for all classes of Midshipmen throughout the fall semester; all faculty and staff will also be trained on diversity, equity, and inclusion."

3. There is an advanced course called Social Inequality—NL450. This course investigates the social and physical constructs of race, gender, and ethnicity in the context of social inequality in America. Particular emphasis is placed on understanding how these constructs, both singly and in combination, affect American society and culture. Smoke?

4. The academy published a slide-presentation of Robin DiAngelo's book *White Fragility*, which summarizes "White privilege" and "White supremacy."

5. The Office of Diversity, Equity, and Inclusion Awards consist of a couple of dozen awards which are bestowed annually based on many types of affinity groups.

6. Naval Academy affinity groups include:

Chinese Culture Club	Midshipmen Caribbean Heritage Club
Filipino American Midshipmen Club	Native American Heritage Club
Italian American Club	Muslim Midshipmen Club
Japanese American Midshipmen Club	National Society of Black Engineers
Korean American Midshipmen Association	Vietnamese Student Association
Latin American Studies Club	Navy Spectrum (group supporting gays, lesbians, and transgenders
Midshipmen Black Studies Club	

No one should presume that pride in one's cultural heritage is not a positive affirmation of self. While professional clubs exist, what is driving this proliferation of groups is based on physical and cultural attributes only. Is there a subculture that is promoting segregation by race and culture? As noted by Victor Davis Hanson in his book *The Dying Citizen*,[284] "Once a man owes more loyalty to his cousin than to a fellow citizen, a Constitutional Republic cannot exist." Do uniforms mean something different now?

Historically, military development takes diverse groups and makes them into highly integrated teams. It is curious that there are affinity groups supporting the sustainment of differences and not unity. Except for summer cruises, there is limited group activity for discussing and advancing common goals in warfare-specialty options which would seem to be much more in line with the mission of the Naval Academy. Is it possible that "yard tribes" are undermining the Naval Academy's sense of being shipmates? Where there's smoke . . .

The USNA Office of Diversity, Equity, and Inclusion (ODEI) directly supports the Naval Academy's Strategic Imperative One: To recruit, admit, and graduate a diverse and talented Brigade of Midshipmen.[285] We understand ODEI is a stand-alone organization that reports directly to the Superintendent.

In addition to the ODEI, there is an additional chain of command within the Brigade Midshipman Command structure referred to as Diversity Peer Educators or DPEs. The intent is to have one DPE in each company and one on each varsity sports team. Although resident in the company/Battalion/Regiment structure of the Brigade, it appears that the DPE's report is external to the normal Chain of Company Commander/Company Officer.

Why is such an organization necessary outside the historical chain of command that has functioned for decades or longer? More important, why is the Naval Academy's Strategic Imperative One to recruit, admit, and graduate a diverse and talented Brigade of Midshipmen? This seems contrary to the Academy's mission which is, *"To develop Midshipmen morally, mentally and physically and to imbue them with the highest ideals of duty, honor and loyalty in order to graduate leaders who are dedicated to a career of naval service . . ."* There is nothing in the Mission about diversity, let alone that it should be Strategic Imperative One. Where there's smoke . . .

From the *Washington Free Beacon* in April 2021,[286] *"Under pressure from student activists, the United States Naval Academy mandated 'diversity, equity, and inclusion' training for students and faculty after complaints of a 'toxic' culture.* But we are told that the Superintendent has stated repeatedly that there is no 'toxic culture' at the Naval Academy. Some of the pressure on the Naval Academy has come from Congress. A bill pushed last year by Rep. Anthony Brown (D- Md.) and Sen. Kirsten Gillibrand (D-NY) called for the collection of demographic data from the Naval Academy's applicants and ranks, specifically focused on racial data. The bill's provisions passed in the 2021 National Defense Authorization Act and now require annual reports on diversity of academy nominations, applications, acceptances, and appointments."

What is driving this asymmetrical focus on diversity at a school whose primary mission is to train combat leaders? Does diversity add to the lethality of the Navy and Marine Corps? What about being judged by the "content of our character?" Where there's smoke . . .

Much of what we have heard comes from Blue & Gold Officers (BGO) in the Naval Academy Information Program. Many of these Officers are

USNA graduates, some retired Navy, and some Marine Corps Officers. There appears to be a level of concern among these volunteers that something is going on at the Academy that they do not understand.

Our BGO data is anecdotal, but after hearing of the possible problems, we were concerned. In light of the BGO organization's unique place in the application process, the Calvert Group sent emails to all BGO Area Coordinators requesting information. From one BGO we were told, "*You are accurate that the email distribution of the Calvert Task Group information was quickly followed by a message by USNA saying that all Area Coordinators were to delete that message—it was not to be forwarded—and that ACs are not allowed to support the CTG mission.*" What are we to think? What is being hidden? What is not anecdotal—but still more disturbing—is that the candidate-interview form no longer requires questions about honor or integrity of the candidate. Where there's smoke . . .

The Naval Academy Board of Visitors (BOV), established by 10 U.S. Code 8468, should be on alert for CRT indoctrination at the Naval Academy. Among its mandates, "*The Board shall inquire into the state of morale and discipline, the curriculum, instruction, physical equipment, fiscal affairs, academic methods, and other matters relating to the Academy . . .*" In November of 2020, the Board's Charter was renewed. The renewing letter from the Chairman of the House Armed Services Committee stated, "*The Chief Management Officer of the Department of Defense (DoD) has determined that renewing the charter for the Board is essential to DoD business operations . . .*"

On 10 September 2021, *The Federalist* reported, "*On January 30, less than two weeks after President Biden's inauguration, Secretary of Defense Lloyd Austin announced he was suspending the Boards of Visitors for West Point, the Air Force Academy, and the Naval Academy . . .*" Meanwhile, on 9 September 2021, CBS News reported that, "*The Biden administration on Wednesday removed 18 appointees named to U.S. military academy boards by Donald Trump in the final months of the Republican president's term in office . . .*" Why were the boards suspended, and why were appointees made by a pro-military president removed? Where there's smoke . . .

Finally, there is the case of a First-Class Midshipman who Naval Academy leadership tried to remove because of anonymous tweets made during summer leave in support of a Police Department that was under online attack during the 2020 summer riots. The Midshipman sued the Academy to remain in the Brigade and graduate with his class. The trial record of that case supplied additional evidence that there is a CRT problem at USNA. A small sampling of that record is enlightening, especially in light of some of the Superintendent's comments about diversity and CRT.

IN THE UNITED STATES DISTRICT COURT FOR THE DISTRICT OF MARYLAND, Civil Action No. 1:20-cv-02839-ELH,[287] the following evidence, in the form of quotes by Vice Admiral Sean Buck, Naval Academy Superintendent, among others, was submitted:

1. On pages 8–9, "The Superintendent, VADM Buck, has made clear through repeated public pronouncements that he unconditionally embraces the structural- and institutional-racism tenets of the BLM, ANTIFA, and Anti-Racist movements and that those tenets will be drilled into the minds of every member of the Brigade of Midshipmen through mandatory training. The Superintendent has also made clear that he intends to root out and separate any midshipman who fails to get 'onboard' with these policies."

2. On pages 24–25, "In a statement issued to the entire Naval Academy community on or about September 9, 2020, the Superintendent applauds the formation of a team under the guidance of the Academy's Chief Diversity Officer that is "developing a midshipman-led, comprehensive plan to identify midshipmen-level shortfalls within our Naval Academy family with the goal of proposing a plan to resolve these issues of privilege, bias, and racial injustice."

3. On page 25, he further notes that "Our Faculty Senate recently passed a resolution with overwhelming support to investigate and address any practices at the Naval Academy that perpetuate systemic racism." The statement is accompanied by a video featuring various Midshipmen who, among other things, discuss their allegiance to Black Lives Matter and make repeated references to "the Movement"— that is, the movement grounded in Critical Race Theory and race

stereotyping, guilting, scapegoating, and conformity of viewpoint. Where there's smoke . . .

And so, the disturbing presence of Woke Smoke comes full circle. From testimony in the U.S. Senate to inconsistent comments by the Naval Academy Superintendent. From outside-source CRT recognition to a new Office of Diversity, Equity, and Inclusion. From BGO concerns to removal of Board of Visitor members. And from a plethora of race-based affinity groups to legal proceedings that shine a disquieting light on an institution which should be held in the highest esteem by the citizens of this country. Even to an objective observer, the dots are connected. The Calvert Task Group remains concerned, vigilant, and ready to serve the Naval Academy, the Navy, and the Constitution of the United States of America, as our oath requires.

★ ★ ★

In Their Own Words: "What Is Critical Race Theory?"

By Phillip J. Keuhlen

Introduction

In more than two decades of military leadership and a further two decades in corporate leadership, I learned the importance of reading and *believing* what my potential adversaries said about their aims and planned methods. During the Cold War, we read Admiral Gorshkov's *Red Star Rising at Sea* and *The Sea Power of the State* to understand what we might face at sea. Today similar reading focuses on the China Aerospace Studies Institute's series, "In Their Own Words: Foreign Military Thought, and Its Translation of China's 'Science of Military Strategy (2013),'" as one hopes today's U.S. military leadership does.

Yet, even though Progressives seceded philosophically from America's founding values a century ago and have conducted lawfare to subvert the Constitution designed to protect those values for over a century,[288] today's U.S. military leadership collectively demonstrates a profound ignorance of a current Progressive dogma mainstay, Critical Race Theory (CRT). They display baffling support for its indoctrination in the U.S. Armed Forces in Diversity, Equity, and Inclusion (DEI) programs.

The underpinning of the U.S. Navy DEI program is its Task Force One Navy Final Report, a deeply flawed effort.[289] It was launched from twin assumptions that the Navy suffers from: systemic racism and that diversity is a benefit to the Navy's military mission. Those assumptions are not supported by evidence, and the use of data and outside scholarship in support of the report's findings and recommendations is flawed by both logical fallacies and by data that is misapplied or misrepresented to support its distorted narrative. Themes that are central to Critical Race Theory and its allied movements that are wholly absent from its precursor document[290] find an active voice in the TF1N Charter and Final Report, confirming a philosophy and a political agenda on the part of those who wrote the Task Force One Navy Charter and steered the enterprise.

Richard Delgado is one of the three founders of Critical Race Theory, the only one still living, and the only one to write broadly about its aims, allies, and methods. His short, seminal work, *Critical Race Theory*[291] (CRT), is pointedly missing from service leaders' recommended reading, which invites the question of intentionality. After all, Delgado positions CRT as a domestic enemy of the U.S. Constitution, stating, "Critical race theory questions the very foundations of the liberal order, including equality theory, legal reasoning, Enlightenment rationalism, and neutral principles of constitutional law."

So, let us delve into the history, intellectual antecedents, objectives, and methods of CRT *in its founding fathers' own words*, as we would any potential foreign adversary's work on strategy. In Sections A-H, all italicized quotes are extracted directly from Richard Delgado's foundational work, *Critical Race Theory*. Bolded passages provide this author's emphasis.

A. "What Is Critical Race Theory?"

> "The critical race theory (CRT) movement is a collection of activists and scholars engaged in studying Diversity, Equity, and Inclusion (DEI) as well as the relationship among race, racism, and power. The movement considers many of the same issues that conventional civil rights and ethnic-study discourses take up but places them in a broader perspective that includes economics, history, setting, group- and self-interest, and emotions and the unconscious. Unlike traditional civil rights discourse, which stresses incrementalism and step-by-step progress, critical race theory questions the very foundations of the liberal order, including equality theory, legal reasoning, Enlightenment rationalism, and neutral principles of constitutional law."

"After the first decade, critical race theory began to splinter and now includes a well-developed Asian-American jurisprudence, a forceful Latino-critical (LatCrit) contingent, a feisty LGBT interest group, and now a Muslim and Arab caucus. Although the groups continue to maintain good relations under the umbrella of critical race theory, each has developed its own body of literature and set of priorities. For example, Latino and Asian scholars study immigration policy, as well as language rights and discrimination based on accent or national origin. A small group of American Indian scholars addresses indigenous peoples' rights, sovereignty, and land claims. They also study historical trauma and its legacy and health consequences, as well as Indian mascots and co-optation of Indian culture. Scholars of Middle Eastern and South Asian background address discrimination against their groups, especially in the aftermath of 9/11."

B. *"Early Origins"*

- 1989 workshop
- Derrick Bell, Alan Freeman, and Richard Delgado

C. *"Relationship to Previous Movements"*

"As the reader will see, critical race theory builds on the insights of two previous movements, critical legal studies, and radical feminism, to both of which it owes a large debt. It also draws from certain European philosophers and theorists, such as Antonio Gramsci, Michael Foucault, and Jacques Derrida, as well as from the American radical tradition exemplified by such figures as Sojourner Truth, Frederick Douglass, W. E. B. Du Bois, Cesar Chavez, Martin Luther King, Jr., and the Black Power and Chicano movements of the sixties and early seventies. From critical legal studies, the group borrowed the idea of legal indeterminacy—the idea that not every legal case has one correct outcome. Instead, one can decide most cases either way, by emphasizing one line of authority over another or interpreting one fact differently from the way one's adversary does. The group also incorporated skepticism of triumphalist history and the insight that favorable precedent, like Brown v. Board of Education, tends to erode over time, cut back by lower-court interpretation, administrative foot dragging, and delay. The group also built on feminism's insights into the relationship between power and the construction of social roles, as well as the unseen, largely invisible collection of patterns and habits that make up patriarchy and other types of domination. From conventional civil rights thought, the movement took a concern for redressing historical wrongs, as well as the insistence that legal and social theory lead to practical consequences. CRT also shared with it a sympathetic understanding of notions of community and group empowerment. From ethnic studies, it took notions such as cultural nationalism, group cohesion, and the need to develop ideas and text centered around each group and its situation."

D. "Principal Figures"

- Originators
  - Bell, Freeman, Delgado

- Major Early Figures
  - Kimberlé Crenshaw, Angela Harris, Cheryl Harris, Charles Lawrence, Mari Matsuda, Patricia Williams

- Asian Scholars
  - Neil Gotanda, Mitu Gulati, Jerry Kang, Eric Yamamoto

- Amerindian Scholar
  - Robert Williams

- Latino Scholars
  - Laura Gomez, Ian Haney Lopez, Kevin Johnson, Gerald Lopez, Margaret Montoya, Juan Perea, Francisco Valdes

- Black Scholars
  - Paul Butler, Devon Carbado, Lani Guinier, Angela Onwuachi-Willig

- Fellow Travelers
  - Andre Cummings, Nancy Levit, Tom Ross, Jean Stefanic, Stephanie Wildman

E. *"Spin-Off Movements"*

"Although CRT began as a movement in the law, it has rapidly spread beyond that discipline. Today, many scholars in the field of education consider themselves critical race theorists who use CRT's ideas to understand issues of school discipline and hierarchy, tracking, affirmative action, high-stakes testing, controversies over curriculum and history, bilingual and multicultural education, and alternative and charter schools."

"They discuss the rise of biological racism in educational theory and practice and urge attention to the resegregation of American schools. Some question the Anglocentric curriculum and charge that many educators apply a 'deficit theory' approach to schooling for minority kids."

"Political scientists ponder voting strategies coined by critical race theorists, while women's studies professors teach about intersectionality—the predicament of women of color and others who sit at the intersection of two or more categories. Ethnic studies courses often include a unit on critical race theory, and American studies departments teach material on critical white studies developed by CRT writers. Sociologists, theologians, and healthcare specialists use critical theory and its ideas. Philosophers incorporate critical race ideas in analyzing issues such as viewpoint discrimination and whether Western philosophy is inherently white in its orientation, values, and methods of reasoning."

"Unlike some academic disciplines, critical race theory contains an activist dimension. It tries not only to understand our social situation but to change it, setting out not only to ascertain how society organizes itself along racial lines and hierarchies, but to transform it for the better."

F. "Basic Tenets of Critical Race Theory"

- "Racism is ordinary, not aberrational"—the usual way of doing business.
  - *"Colorblind" policies "can thus remedy only the most blatant forms of discrimination"*

- System of white over color ascendancy (i.e., systemic racism) serves important purposes for dominant group
  - *"Interest convergence"—"Because racism advances the interests of both white elites (materially) and working-class whites (psychically), large segments of society have little incentive to eradicate it."*
  - Race and races are a *"social construction"* that *"correspond to no biological or genetic reality."*
    - Categorizations based on race are invented, fluid, and manipulated
    - Differential racialization—changing stereotypes based on factors such as economic, political, and security needs
  - *"Intersectionality and anti-essentialism"*—no one has a unitary identity—*"everyone has potentially conflicting, overlapping identities, loyalties, and allegiances."*
  - *"Unique voice of color"*—conflicts with anti-essentialism: only persons of color can speak with competence about race and racism.

G. "Hallmark Critical Race Theory Themes"

- *"Interest Convergence, Material Determinism, and Racial Realism"*
  - *"Racism is a means by which society allocates privilege and status"*

- *"Revisionist History"*[292]
  - Replacement of majoritarian interpretations

- Materialists vs Idealists in CRT Activism
  * Materialist focus: unions, immigration, job outsourcing
  * Idealist focus: speech codes, media representation, diversity seminars

- *"Critique of Liberalism"*
  - Rejection of color-blindness and neutral principles of constitutional law (equality of treatment and opportunity)
    - *"Aggressive color-conscious efforts"*

- *"Structural Determinism"*
  - System by reason of structure and vocabulary is ill equipped to address certain types of wrongs.
    - *"Tools of Thought"* in law are *"liberal/capitalist"* basic structure
      * Do not address CRT issues: intersectionality, interest convergence, microaggressions, anti-essentialism, hegemony, hate speech, language rights, Black-White binary, jury nullification.
    - *"Empathic Fallacy"*
      * *"Hate speech is simply not perceived as such"*
    - *"Serving Two Masters"*—Conflicts of Interest in Law Reform
      * Do activist attorneys or legislators serve the best interests of their immediate client/constituents or pursue social change?
    - *"Race Remedies Law"*—Civil Rights Law and Enforcement as Impeding Change
      * Moderates pace of changes to satisfy all parties.
      * When radical change is necessary, elites allot impacts to Minorities (stigma) or working-class Whites (displacement)

H. *"A Critical Race Agenda for the New Century"*

- "Erase Barriers to upward mobility for minority populations, especially tests and standards for merit"

- "Economic Boycotts aimed at increasing minority representation in the media" and counter (cancel) media that offends minorities.

- "Rectifying racism in policing and criminal justice"—"better chance of going to college than to jail"

- "Sentencing reform and attention to postconviction consequences"—felony disenfranchisement

- "New immigration policies"—"freer flow of labor and capital" while maintaining strength of unions

- "Cease requiring assimilation"[293]

- "Economic democracy" (Economic democracy [sometimes called a Democratic Economy] is a socioeconomic philosophy that proposes to shift ownership and decision-making power from corporate shareholders and corporate managers (such as a board of directors) to a larger group of public stakeholders that includes workers, consumers, suppliers, communities, and the broader public.)[294]

- "Assure, through appropriate legislation and other structural measures, that the reforms cannot easily be undone."

Conclusion

The evidence is inescapable: much of today's military leadership has embraced Critical Race Theory,[295] is preparing their leadership academies to indoctrinate it,[296] and is obfuscating those two facts in testimony to the Congress.[297] Previous work has dwelt on the question of whether Navy leadership understands the political philosophy embedded in its TF1N Final Report or the political intent inherent in its Diversity, Equity, and Inclusion (DEI) programs. Two years later, motivation doesn't matter. American citizens need to understand that their Armed Forces Leadership has embraced and doubled down on DEI.[298] The current nominee for the Chairman of the Joint Chiefs of Staff is either more interested in the racial balance of the Armed Forces than its combat effectiveness or believes that combat effectiveness can be achieved through racial balancing. Try running the latter perspective through the management or fan base of any major professional or college sports program (including the Service Academies' sports programs) to see how that logic flies in organizations laser-focused on winning!

It is long past time for American citizens to speak loudly through their representatives to block the career advancement of activist federal appointees, General, and Flag Officers and purge Critical Race Theory indoctrination from military and Federal Service. Per the DOJ, "Title VI, 42 U.S.C. § 2000d *et seq.*, was enacted as part of the landmark Civil Rights Act of 1964. It prohibits discrimination on the basis of race, color, and national origin in programs and activities receiving federal financial assistance." Some argue that Title VI applies only to organizations receiving federal assistance, not to the Congressionally funded federal government itself. But eight decades ago, the Armed Forces dismantled discrimination based on race and virtually all Federal departments and agencies have policies that mirror Title VI in most regards. Based on recent public testimony,[299] Armed Forces leadership apparently doesn't see that as unequivocally the case with respect to DOD. This is an egregious situation the Congress should promptly address. Congressmen and Senators should start by demanding formal responses from SECDEF, Service Secretaries, and each service's senior uniformed officer to the questions:

- "Would you support legislation to clearly extend Title VI, 42 U.S.C. § 2000d *et seq.* to apply to the organization you lead, without exception?"

- "In the interim, will you pledge ***under oath*** to strictly enforce a ban on ***any*** consideration of race in military personnel management such as accessions, assignments, promotions, etc.?"

The Senate should put DOD in the position in every public nomination hearing of defending an indefensible position—to say that nearly eight decades after President Truman's 1948 Executive Order 9981 began desegregation/integration in the military and seven decades after the unanimous 1954 decision in *Brown v. Board of Education*, and more than half a century after passage of the landmark 1964 **Civil Rights Act,** DOD reserves the right to discriminate against military personnel based on race.[300] Such claims cannot be condoned and should not be tolerated.

Senators should vote against confirming those who do not clearly answer in the affirmative.

CRT and its DEI cadres aim to undo and remake America. Much as we read Marx and Mao in the late '60s at USNA, it is wise to educate, *not inculcate*, about this pernicious philosophy and the tenets that it seeks to embed into law such that they cannot be undone:

- Questions the very foundations of the liberal order, including
  - Equality theory,
  - Legal reasoning,
  - Enlightenment rationalism, and
  - Neutral principles of constitutional law.
- Sees Racism as ordinary, not aberrational.
- Rejects color-blindness and neutral principles of constitutional law (equality of treatment and opportunity) in favor of "Aggressive color-conscious efforts."
- Rejects "moral and legal rights."
- Focuses on racially distributed "equality of results (Equity)" rather than "equality of opportunity."
- Would eliminate merit as the foundation of personnel management.
- Abandons our founding emphasis on requiring assimilation of immigrants.
- Economic democracy.

Enough is enough. Silence is consent!

★ ★ ★

In Their Own Words: "What is Critical Race Theory?" Real Clear Defense, 8 August 2023

Generational Responsibility

By Tom Burbage

Look up the word "generations" today, and there may be as many as there are emerging genders. But the reality of generational responsibility lies in the world one generation leaves behind for the next or the one after that. Historically, adult conversations have recognized succeeding generations "stand on the shoulders" of the ones that came before us. We are an "apprentice based" society, with an inherent responsibility to mentor those coming behind. The world we leave should be better than the world we found. For those who came before, it was always worth fighting for and, in many cases, dying for. Not so today. Have we abandoned that generational responsibility?

It matters not whether you believe in the catchy terms of Baby Boomers, Millennials, Generation X, Generation Z or even, respectfully, the Greatest Generation. While each may have their unique qualities or motivations, they all should feel a biological need to provide a legacy for their own next generation. That idealistic concept may soon become an unachievable dream.

We are, and have always been, a Constitutional Republic. We elect people to represent us in our Congress, both Senate and House. We elect a President. We depend on honest and open journalistic freedom to shape our opinions and, in a perfect world, we hope to live up to our generational responsibility. Where and when did that hope come off the rails? It's hard to say, but it has been going on for many years. Somehow the forces of evil have invaded and taken root in our historic American DNA as the "beacon on the hill." The door has been opened for an insidious infection of a Marxist ideology that may threaten any hope of achieving that generational responsibility.

In the end, it comes down to a matter of trust, or lack of it, in our foundational institutions. We trusted that our elected leaders would take an oath to support and defend our Constitution. We trusted that our election process would maintain purity to elect leaders that would truly represent our interests, as a nation. We trusted in our sacred and apolitical military leaders. We trusted in our school systems. Or . . . we used to.

So "Trust" has become yet another word that is hard to define.

Our forefathers recognized the human frailties of 'trust' and created the three pillars of our form of government, each meant to control that elusive element. They were very smart men, and their legacy was prescient for our nation. Their future would be marked by success or death by hanging, so they truly had skin in the game. Our Constitution remains very relevant today. They recognized the concept of generational responsibility.

Perhaps taking a closer look at the state of our nation should be done from their perspective.

Our open Southern border allows the unimpeded influx of migrants, enriching the cartels and imposing severe financial strains on local communities, a perverse inversion of our capitalist economy. Overwhelming our welfare system is a key element of the Marxist playbook, not our Constitution. Doing it at the expense of our veterans and our homeless is a further afront to our founding principles. The consequences of the profusion of dangerous drugs across the open border, coupled with the transportation of undocumented migrant elements throughout the entire nation, is counter to any sense of national security, one of our Constitutional mandates.

There is a slow rot occurring with the continual elimination of criminal justice. The allowance of an unexplained tolerance of smash-and-grab tactics, the no-cash bail policy, and the "Defund the Police" initiatives make no sense to most Americans. Making common crime an acceptable consequence is fundamental to the unwelcome move toward Socialism. We have somehow ingrained a justice system that not only supports these new initiatives but is now resulting in a two-tiered system steeped in political partisanship.

This is not what the Founding Fathers envisioned, and it's antithetical to our Constitution. We are rapidly moving toward a world where our grandchildren may one day speak Mandarin.

Most critical to our community of veterans is the divisive ideology that began with liberal universities but has now permeated our public-school systems. The resultant poisoning of the military-recruiting process has set back our ability to attract the next generation of military leaders. This infection is now permeating the senior leadership institutions of our services and our Service Academies. Inspiring ordinary people to do extraordinary things is no longer the mantra of our military leadership. This loss of focus within the military may be the single largest threat to sustaining our leadership within the global community.

We live in a world of good vs evil. Words matter, and they mean very different things to different people. Often, they are carefully manipulated to do exactly that. Our nation's legacy of strength and leadership has historically been based on our sense of God, family, and patriotism and are codified in our Constitution. All three of those are under assault. The erosion of our strength and leadership has been well recognized by both our allies and our enemies. Nature abhors a vacuum, and the same can be said about our leadership on the world stage.

There are arguably many faults in our journey to this point in our evolution as the strongest nation on the planet. But there are also many unarguable strengths. No society is perfect. Societal suicide is a byproduct of abandoning founding principles and succumbing to human frailties like greed. We are seeing that today.

Our Founding Fathers lived up to their generational responsibility. Our Greatest Generation fought for our freedoms in World Wars 1 and 2, Korea, and Vietnam. They endured sacrifices and challenges that our generations will never know and, apparently, will never appreciate. They lived up to their generational responsibilities.

So . . . how about us? Are we passively leaving it to our kids and their kids to dig themselves out of this ditch?

There aren't many options. We can be ignorant or apathetic and take no responsibility. "Not our problem" is a common refrain for those with a misplaced trust in the institutions that manage the state of our nation. Or . . .

we can be activists, change agents, and vocal advocates to help recapture the basic freedoms we enjoyed that came at a steep cost in human and national treasure from those that came before us. Or . . . we can silently salute the freedoms we have enjoyed and welcome those that come behind to a new world we never had to experience.

Which will it be? It all comes down to two words: Generational Responsibility.

★ ★ ★

Generational Responsibility, Armed Forces Press (AFP), 31 August 2023

No Institution Is Safe: Thought Control in the Military

BY P.J. KEUHLEN AND SCOTT STURMAN

There is a misunderstanding that brainwashing, a technique of mental and psychological reprogramming conducted in an environment of ideological totalism,[301] is irresistible and permanent. However, social isolation, sensory and sleep deprivation, torture, and psychological manipulation in a dystopian environment do not transform most subjects into passive automatons that are amenable to any and all suggestions.

A far more successful system of thought control and persuasion is described by founders of Critical Race Theory (CRT), who far better understood the psychological motivations required to instill long-lasting and uncompromising cognitive alterations. Their genius was to disguise this obscure, destructive Marxist philosophy by identifying the operational component of CRT with three benign words that appeal to fairness and the fellowship of the human race—diversity, equity, and inclusion (DEI).

Brainwashing and DEI share a spectrum[302] of similar mind-altering practices, including the strict control of word definitions and speech patterns, the emphasis on confession without absolution, the forfeiture of individual identity to the group, and the labeling of detractors in absolute, pejorative terms. But unlike the brainwashing techniques employed in the Chinese prison camps of the 1950s, DEI offers its subjects a sense of belonging and a path to the self-defined moral high ground that has captured the will of millions who are willing to devote their lives with near-religious fervor to the transformation of the world's institutions.

In 1950, the journalist and CIA operative Eduard Hunter introduced and glamorized the term *brainwashing*[303] to describe the coercive methods of mind control the Chinese Communists employed against U.S. POWs during the Korean War. His sensational claims of an irresistible form of indoctrination that rendered its subjects intellectually placid and remodeled evinced parallels to the fictional works of *Brave New World* and *1984*. The movie *The Manchurian Candidate* led the public to speculate that there were those among us who could be activated by a simple word or deed to metamorphose from an everyday citizen to an active Communist agent.

The psychiatrist Robert Jay Lifton repudiated many of Hunter's claims, citing evidence from his extensive interviews of both military and civilian prisoners who were the targets of intensive, programmed-thought reform. Lifton stated that the process could be resisted, that its implementation was systemic, and that the methods were not exclusive to the Chinese. Supporting his claim was that only 21[304] of 22,000 U.S. POWs refused repatriation, while the remainder, despite receiving comprehensive mental reprogramming, elected to return home.

Lifton published his findings in 1961 in the book *Thought Reform and the Psychology of Totalism: A Study of "Brainwashing" in China*. He listed eight elements[305] that form the basis for intimidatory mind programming that share similar psychological objectives with DEI. Communication is highly controlled, with the reduction of language to easily remembered clichés in a system where subjects do not realize they are being manipulated. Purity of thought is a requisite, and it is defined in a good vs. evil dialectic that considers opposing doctrines as illegitimate. Ideology is sacred, and one's character must be shaped to fit the template. Those who stray from the doctrine must confess lapses, while non-repentant detractors have no authority to express contrary opinions.

In a 2014 interview,[306] Dr. Lifton reiterated that the term "brainwashing" was a misleading construct and that he preferred the terms "thought reform" or "mind control." "Brainwashing" imputes an all-or-nothing phenomenon and does not account for different types or levels of persuasion. He provided two examples applicable to the political and academic setting that he described as "more gentle expressions of totalism." The politician can be compelled to confess for failing to adhere to political orthodoxy, and the student can be subjected to psychological coercion for failing to attain proper achievement, depending on the ideas promulgated by one's teachers.

For thirteen years, impressionable[307] K-12 students are bombarded with relentless propaganda promoted by teachers who interact with them as trusted adult authority figures. The two largest teachers' unions in the United States, the National Education Association (NEA) and the American Federation of Teachers (AFT), staunchly support DEI, and its member teachers could be described as its disciples. The NEA's three million educators and retired members are pledged to promote inclusivity[308] and racial justice[309]—both politically charged terms drawn from the core of Marxist critical theories. The smaller AFT includes 80,000 educators and 250,000 retired members, but the organization's DEI and racial-justice resolutions[310] read more like the Occupy Wall Street Manifesto[311] than a pledge to provide the highest quality of merit-based education.

By the time a cadet or midshipman enters a United States military academy, most of them have been subjected to the "more gentle expressions of totalism" from grammar school through high school. DEI indoctrination methods brilliantly lull both students and parents into complacency by branding it as a philosophy that embraces equal opportunity and inclusiveness rather than a nihilistic, radical doctrine that advocates anti-capitalism, anti-free speech, and the primacy of the state over the individual. From the guileless first-grader to the high-school senior, the student's exposure to DEI is promotional, which explains the ease of its "long march to the institutions"[312] envisioned by the father of the new left, Herbert Marcuse.[313]

Acceptance to a military academy represents a crossroads where adolescents and young adults anticipate the experience of an intensive, traditional military environment, where they will have the opportunity to live in a setting[314] where behavioral expectations are centered on ability, unity, and service. In theory, the academies present a stark contrast to civilian institutions, where pervasive DEI programs promote individual identity defined by phenotype and sexual orientation, a culture of power structures and victimhood, and the idea that guilt and genetics are inseparable.

For more than a decade, since the introduction of President Obama's Executive Order 13583, U.S. military academy administrations have deemphasized merit[315] and embedded DEI programs[316] into the fabric of academy life, describing them as a military necessity and on par with academic performance. The faculty and staff at the United States Naval Academy (USNA) advance the principles and practices of critical pedagogy[317] by inserting critical-theory principles into both the social sciences and STEM curricula. A summary of Critical Race Training[318] at the USNA and its DEI informational page[319] on the academy's website describes an odd mixture of an unabated institutional commitment to inclusiveness and diversity, while simultaneously endorsing segregated affinity groups based on ethnicity, sex, and race. The most recent USNA DEI conference excluded all participants who did not receive an invitation—an indication of a lack of tolerance for competing ideas and the advancement of sanctioned opinions viewed through the lens of racial identity.

The United States Military Academy's (USMA) 2023 DEI Conference revealed an academy administration that has faithfully fulfilled Marcuse's bidding by enlisting the corps of cadets to serve as soldiers in support of DEI. USMA officials proclaim DEI's indisputable benefit to the military, citing evidence[320] from the financial-services and management-consulting sectors, but these under-powered studies were conducted in a limited, non-military setting. Comprehensive studies[321] from Harvard and Tel Aviv Universities of 800 companies spanning 30 years contradict this view and demonstrate that DEI programs frequently do not change attitudes and often aggravate racial biases and hostilities.[322]

Testimony[323] in July 2023 by the superintendents for all three major Service Academies at the House Armed Services Committee demonstrated a uniformity of opinion often delivered in a talking-point format fraught with clichés. Their vigorous commitment to DEI unveiled a degree of smugness, not unlike the testimony exhibited by three high-profile[324] Ivy League university presidents, whose comments portrayed them as ideologues defending tenuous positions. High-ranking generals, who defend programs that promote racial discrimination and disregard merit as the preeminent predictor of student success, do little to gain the public's trust as guardians of the next generation of military leaders.

All of the U.S. Service Academies have adopted efficient versions of thought-reform programs described by Lifton. Military-academy campuses represent a milieu that is spatially and socially isolated from the general public. The atmosphere is rigidly hierarchical, both in terms of professor-student and officer-subordinate relationships. For four years, academies function as a Petri dish, where radical ideas can be imposed under the guise of military and academic training. Late-stage adolescents and young adults are highly impressionable and vulnerable[325] to external stimuli when subjected to subtle-but-comprehensive propaganda.

The psychological pressure, potential abuse of power and intimidation, and specter of anonymous accusations are foreign to non-totalitarian military organizations. Yet DEI, which permeates the lives of Air Force Academy (USAFA) cadets and Naval Academy midshipmen, employs these techniques. At both USAFA and USNA, cadet political officers,[326] who wear a distinctive armband, are installed in each squadron or company and tasked to report DEI-related information outside the chain of command to the academy's Chief of DEI. White-male cadets at USAFA are subjected to harassment in the classroom and forced to explain their White privilege.[327] A Cornell-trained, civilian professor of economics identified a White male cadet not by name but as "White boy #2."[328] She informed the class that she was inclined to do so, since all White people look alike.

The window of opportunity to obtrude years of DEI indoctrination and train officers in accordance with the highest standards of military science has been squandered. Cadets and midshipmen hoping for a reprieve from DEI propaganda become disheartened, and their classmates, who support DEI dogma, are emboldened by the affirmation granted them by professors and members of the military training department. Congruent opinions between superiors and students in the academy setting offer distinct rewards—enhanced prospects for promotion and academic and professional advancement. The academies have become a reliable source of pro-DEI military officers, and, although they commission[329] less than 20% of the officer corps, these graduates command a disproportionate influence within the Department of Defense.

Just as excess death rates[330] provide a measure of general public health, the U.S. military recruitment[331] crisis serves as a litmus test of the health of the United States Armed Forces. Despite overwhelming support of the DEI culture[332] by members of the Department of Defense, Congress, and the White House, a career of service to the country is no longer appealing to many young Americans, particularly those from families whose service[333] is generational. Wokeness,[334] affirmative action,[335] CRT, and DEI[336] are not recognized by the public as a panacea for military readiness and power,[337] but as a means to incapacitate and infect the armed services with Marxist propaganda. 40% of the admirals and generals[338] whose promotions were delayed have made public statements supporting DEI. Once a critical mass[339] of officers reaches 30%, DEI will be self-sustaining, and it will take years to reverse the trend.

The effectiveness and permanency of classic brainwashing programs never materialized. They lacked the ability to subtly influence and manipulate young minds over long periods of time. The laws of physics illustrate that the application of a small force over an extended period of time can produce large changes of momentum to an object, provided the time is sufficiently long. DEI thought reform draws on this analogy. DEI propaganda succeeded where traditional Marxist thought revision failed by understanding the human psyche's vulnerability to incremental influence over long periods at the crucial periods of psychological development. DEI's ability to persuade large populations to voluntarily act in a self-destructive manner marks it as a triumph of Marxist brainwashing.

Cadets and midshipmen face a daunting task—how to resist DEI indoctrination that applies thought reform through coercive, yet often subtle, psychological techniques. They are the targets of unethical and subconscious manipulation that are purposely directed at them without informed consent. The systematic intellectual and behavioral seduction of susceptible college-age students without their express knowledge for the purpose to achieve a contrived outcome exposes them to the dangers of medical experimentation.[340]

Adolescents and young adults attending U.S. Service Academies are expected to take orders with little reservation, but they are also patients, who are entitled to the protections afforded by the Nuremberg Code.[341] Cadets and midshipmen were required to receive Covid-19 mandatory vaccinations without proper informed consent. Now they endure the insidious processes of thought reform that are psychologically invasive and often result in long-lasting effects. Their right to understand the fully disclosed risks and benefits of these intrusions—and the option to refuse them—protect them from these abuses.

Patients living in highly structured organizations where "proof by authority" doctrines are preeminent often lack the safeguards that ensure their basic human and medical rights. Military medical professionals, particularly psychologists and psychiatrists, have the obligation to publicize unethical medical practices that are conducted without the patient's voluntary consent. The need to serve as patient advocates and to identify and resist DEI indoctrination protocols represent the standard of care.

Members of the medical community are the select few in the present-day military who can offer safe harbor to those exposed to unremitting propaganda. DEI is a scourge of the Armed Forces, and commanders should cultivate relationships with military medical and legal professionals to channel their efforts to challenge DEI programs at every level.

★ ★ ★

No Institution Is Safe: Thought Control in the Military, (Joint author w/ Scott Sturman), Brownstone Institute, 21 December 2023

Where Do We Go from Here?

Is there a path forward? Can this situation be turned around, and can we regain or maintain our status as the top military power in the world? Can we regain the trust of our service members, our nation, and the world? We think so. That is why our organization exists. We have to make the situation known to the leadership but, more importantly, to the nation that elects the officials who make the budgetary, end-strength, acquisition, and, especially, the policy decisions that provide for military operations.

We often talk about "situation awareness" when addressing military operations. Awareness is lacking in the nation in general, partly due to extremely partisan media that refuse to tell the truth. Rather than transparency, we have concealment. The nation is living in darkness. Our purpose is to shine a light on the situation, particularly with respect to the Armed Forces. We join our voices with those also seeking sanity and the future of the nation, such as the flag and general officers who recently spoke out on this issue (Flagofficers4america.com).

We believe that, with enough publicity, we, along with others, can stem the tide of decline and perhaps even turn it to put us back on the path of greatness. Through contact with our elected officials, engagement in political campaigns, and joining our voices with others reaching out to the public, we hope to bring the awareness of the detrimental policies out into the open and help instigate a turnaround. The policies dragging down our military and our nation are touted as "feel good," "nice," "concerned for all," etc., while hiding the true negative effects on our status as a world power. They also harm the population at large under the guise of "helping everyone." We hope that this message will be transmitted far and wide and renew our nation's interest in the Constitution, socio-economic policies, real science, and engagement by "we the people" rather than the withdrawal that has been going on for several decades.

★ ★ ★

It's Time to Choose

By Tom Burbage

I THINK THAT WORDS MAY BE THE NEW BATTLEFIELD, as they have subtle meaning that can motivate the new "armies" that seem to be changing the definition of "enemies." The best example is the difference between Awake and Awoke. People are either Awake or Awoke, and it is interesting that these two words, separated by a single letter in the alphabet, may define the new boundaries of combat. If you declare you are neutral, you are by definition "Awoke"... willing to follow the lemmings over the cliff. If you are "Awake," you recognize that the lemmings are a false idol.

We all knew people from the "Greatest Generation" who gave everything to keep the American experiment alive. Many were our parents and our role models, and we often recalled their contribution as a key part of our development as young adults. Over a half century after World War II, we all knew friends who were killed or caught up in 9-11 and saw our "next-generation" volunteer to jump into the fray. After 9-11, our streets were lined with American flags, and young patriots flooded our recruiting stations because it was a visible threat to our nation and our way of life. More Americans were killed on 9-11 than were killed at Pearl Harbor, an event that brought us into a World War. Since those two major events, it is ironic that we have endured no national threat. Some incidents require military responses, yes, but our shores and families have never been threatened, until now. Our generation experienced a unique slice in American History, dealing with two very different threats to our nation. Both influenced our lives in many ways. We overcame both and lived on to enjoy the freedoms we often take for granted. Our generation now looks at the state of our nation through the same lens. That might be a big mistake.

We are now facing a much more insidious, hard-to-define, internal threat with the same potential outcome to destroy our freedoms and way of life. This new insurgency is what our enemies have long predicted—that it is possible to defeat the mighty U.S. without firing a shot. The counterculture has made quiet inroads over our lifetime, expanding without resistance through social media and liberal schooling. Our generation was a contributing factor through our permissiveness that allowed merit to be secondary to feelings, a concept that also facilitated the current dilemma we face.

The enemy today is not "over there" and does not wear a strange uniform or follow the general rules of conflict as we have always known them. There is still plenty of that going on in the world today that we need to be concerned about but . . . . we are now also confronted with a new-ideology focus which seems to act like an insidious cancer from within. Like that disease, the problem often goes undetected but eventually starts to influence critical elements of life. Detected early, it can often be resolved or at least delayed, providing a few more years to enjoy life. Like many of our generation, when faced with a life-threatening medical diagnosis that we don't understand, we search for a source of knowledge that can "make the miracle happen."

It rarely does, particularly when we ignore the symptoms and write them off to "the aging process."

Today, our social fabric has been infiltrated with malware, to steal a term from the age of laptops. Unfortunately, the toxic combination of unconstrained teaching paradigms that none of our generation would accept but which we allowed to infest our schools, plus a political agenda that put a puppet up who could spout the liberal agenda, is just as threatening as any conflict we have faced in our history. The most perplexing question is . . . why does anyone want to go down this path? One of the laws of nature has always been that "the forces of evil are fully funded," a factor that needs to be fully investigated.

Today, one of our great challenges is the recruitment of the next generation of military leaders and our volunteer forces. In an environment of radical learning based on Marxist principles of disunity and hatred, can we expect our best and brightest younger generation to still carry that patriotic DNA that Americans have always had? If not, can we restore it in our Service Academies and other accession processes? Based on human nature, we may have a very reduced chance to motivate our society to respond in defense of our national interests.

So, how do we fight this with incomplete knowledge and a divided nation?

Storm the beach? . . . What beach?

Hand-to-hand combat? Whose hand?

Motivate our young patriots and send them into battle . . . Where is the battle?

Who is the enemy? Is it he/she or some other self-identifying pronoun?

We can spend all of our time chasing political ghosts, or we can pull our heads out of the sand and recognize we are in a very dangerous time for our Constitutional Republic. If we don't identify the threat, stop the insanity over the rejection of what made us the greatest nation in the world, and rise up, we are, in fact, giving into the cancer and hoping for a peaceful ending to a life well lived.

And . . . hoping that our kids and grandkids survive.

Or . . . we can come off the golf course and stand up for our lifelong contribution to who we are as a nation. Too many made the ultimate sacrifice to surrender our national identity to be less than the best. Many of us are enjoying the fruits of the sacrifices of the young patriots who joined the fight voluntarily after 9/11. Would those individuals want their life back to enjoy the lives we have today? Would that large community of "young patriots" with disabling injuries feel the same? No question they would.

Would the same group watch us give our country away to something very less than what they gave their commitment to?

We have an obligation, as our oath so eloquently defined, "to support and defend" the Constitution of the United States. Never forget that oath, even—and especially when—everyone around us is trying to ignore it. It's time to choose what side we're on.

★ ★ ★

The Ultimate Conundrum

By Tom Burbage

"We are a nation of laws, not of men." Thomas Jefferson, Founding Father.

"When our nation's laws are flouted by the political leadership, it is only a matter of time until the populace shows the same disrespect for law and order," John Adams, Founding Father.

They seemed to be wise men predicting our future some 200 years later.

What are we witnessing today? Who are we?

Most of us are in the business of bringing back the good vision of America . . . How do we do that? Time is of the essence—so, what must we do in the next day, the next month, the next year? When will things change? Not before you demand it. Is there a sense of urgency, or not? These questions will define who America is going forward.

A lot has happened in just the last year in our country and around the world. Incredibly, our memory, shaped by the misrepresentations we hear today in our media, drives a forgetful phenomenon, and, unfortunately, these reinforce the whole cycle. Events that would have changed public opinion today are quickly forgotten.

The debacle called an "exit from Afghanistan" was written in the disappearing ink of partisan politics. The death of thirteen Americans and the many left behind are an asterisk on the accounting of a twenty-year sacrifice of blood and treasure. Unless you are a veteran or come from a family of veterans, you cannot appreciate the billions of dollars of U.S. military equipment left behind, making the Taliban a legitimate military force. The price we paid over the last twenty years did not deserve this ending.

Who can deny the corruption related to the 2020 election process or Hunter's laptop, both the victim of disappearing ink?

Over the last two years, many of the things we enjoyed as a freedom-loving country became compromised, and many of the metrics that made America the envy of the world disappeared.

America is showing a surprisingly sudden tolerance for crime, including violent crime—*plus* an unimaginable commitment to opening our borders to the unfiltered incursion of illegal immigrants. More critically, the infusion of deadly fentanyl and other drugs into our most vulnerable societal elements underpins the domestic enemy our forefathers warned us of but which we have never adequately defined. It is here and it is now. And we are unprepared to defeat it.

Is the majority of our nation driving our agenda? I think not.

On the downside, unchecked inflation, food shortages, and, probably most important, an attempt to fatally alter energy in a world that needs it badly, has taken us from global leadership to court jester on the world stage. On the upside, good cops, acts of kindness, and military competence are still strengths, but you would never know it if you follow "the news."

The very vocal Minority is overriding our Democratic process.

Where are all these silent people? Too comfortable to engage? Maybe, but probably not. Uninformed? Possibly. Willing to engage? We'll see. Dependent on their Representatives? Probably so, but there needs to be a trust-but-verify dimension.

Confidence in our country and its leadership has radically diminished, at least among the informed and concerned. The only hope remaining is that the backbone of our democracy is strong enough to right the ship and recover the fundamental precepts that made this country the beacon of freedom for the rest of the world.

So . . . when will things change?

When you demand it and when you individually decide the current path will destroy the future for our children and grandchildren. They will inherit either your strong DNA of defense of our liberties or your weak DNA that follows the lemmings over the cliff. The future of our nation is truly in your hands, maybe for the first time in our history.

Democracy is a participation sport. There was never a greater need to participate than now.

★ ★ ★

National Service—A Powerful Woke Antidote

By Bill McCauley

A POWERFUL WAY TO ADDRESS the accelerating breakdown of civil society in America is to seriously consider a National Service program for all 18-to-25-year-olds, as proven successful in other free countries. By instituting such a program, America could begin the long process of bringing people of all races, creeds, and genders back together again with a unified vision of equality of opportunity for all, as envisioned by our Founding Fathers.

Origins of Wokeness

Both the vision and the reality of equality of opportunity for all has been steadily eroded since LBJ's introduction of the various programs of the Great Society in 1964, which served as the foundation for America's growing welfare state. Although well intentioned, the Great Society's fatal flaw was that it guaranteed equality of outcomes (Equity) at the expense of equality of opportunities, thereby reducing individual initiative and incentive. The Great Society paved the way for the current Great Divide. Covid was a convenient excuse for the Progressives to create the *Mother of All Great Societies*, with an explosion of ill-conceived bailouts directed to broad swaths of American society, and quickly adding trillions to our national debt. A powerful complement to such disincentives to work is the proliferation of DEI programs throughout government, academia, and the military. DEI is code for affirmative action, which is illegal in nine states. On June 29, 2023, the Supreme Court ruled against the use of race in college admissions. Until merit and excellence are reinstated as the key determinants for advancement,

America will not recover from the insidious and deleterious effects of these bankrupt policies. Without a renewed emphasis on thoughtful incentive programs, America's continued decline is assured. And, if not addressed soon, it might be too late to put America back together again.

Possible Solution
With a National Service program, kids from lower-income communities would be mixed in with kids from the country clubs during their service-corps boot camps. This melting-pot experience is a powerful and memorable indoctrination, and it would serve to peel back the vestiges of inequality of opportunity as, in this microcosm of society, all privileges would be earned based on merit, performance, and excellence. The program would be all about being part of a team among equals.

During their training, these kids would be taught that freedom is not free, nor is the service coming from our government. Once they have served in one of a variety of service-corps programs around the country, they will appreciate the key tenet of this program—i.e., *quid pro quo*. And they will observe the benefits of meritocracy and excellence as the key determinants of advancement. If they want to be "served" by any of the myriad social programs of our government, they must first show that they have "served" their country. Importantly, this does not have to be military service.

Reference. Creation of such a National Service program is a tall order, especially in today's Woke society, but it is time to start the process, as Major General John Borling, a former Vietnam POW, has done with: www.sosamerica.org

★ ★ ★

The Constitution, the Officer's Oath, Core Values, and Admirals

By Brent Ramsey

The Constitution is the supreme law of the land. The oath that I took upon commissioning was a solemn vow of allegiance to that Constitution and says in part, "I, XXX, do solemnly swear (or affirm) that I will support and defend the Constitution of the United States against all enemies, foreign and domestic; that I will bear true faith and allegiance to the same; . . . So help me God." That oath that I took 54 years ago still binds me to my country.

At the Navy website are core Navy documents.[342]

Sailor's Creed

I am a United States sailor. I will support and defend the Constitution of the United States of America, and I will obey the orders of those appointed over me. I represent the fighting spirit of the Navy and those who have gone before me to defend freedom and democracy around the world. I proudly serve my country's Navy combat team with Honor, Courage, and Commitment. I am committed to excellence and the fair treatment of all.

Navy Core Values

- Honor: I am accountable for my professional and personal behavior. I will be mindful of the privilege I have to serve my fellow Americans.

- Courage: Courage is the value that gives me the moral and mental strength to do what is right, with confidence and resolution, even in the face of temptation or adversity.

- Commitment: The day-to-day duty of every man and woman in the Department of the Navy is to join together as a team to improve the quality of our work, our people, and ourselves.

The Expanded Honor Section says:

"... we will: Conduct ourselves in the highest ethical manner in all relationships with peers, superiors, and subordinates; Abide by an uncompromising code of integrity; Illegal or improper behavior or even the appearance of such behavior will not be tolerated."

The Constitution, the Oath, the Sailor's Creed, and Navy Core Values tell you everything you need to know about being a good Sailor. The stress is repeatedly on the concept of honor. Everyone understands the importance of honor! These fundamental values and truths obviate the need for introducing divisive social-justice ideas into our Navy.

Questions for the Admirals

Today there are approximately 224 Admirals in the Navy out of an officer corps of 56,000, or about .4%. It is extraordinarily hard to make Admiral, with a tiny fraction of a percent achieving that level. Even to become an officer in the Navy is highly selective, with only a small percent of those applying to be an officer candidate either through the United States Naval Academy, NROTC, or OCS being accepted. To make Admiral is an extraordinary accomplishment reserved for the most highly qualified, most experienced, most dedicated, most honorable leaders. It is to this group of extraordinary patriots that I speak.

Since 2011, the Navy has been ordered by Presidents Obama and Biden to implement Diversity, Equity, and Inclusion (DEI). Why is this necessary, when the overwhelming values of the Navy are honor, integrity, teamwork, and ethical behavior, which the Navy drums into you from the day you take the oath until you are mustered out of the service? Yet, what we find is that the Navy has completely embraced implementing divisive DEI programs into the entire force.

So, here are some respectful but urgent questions for the Navy's Admirals:

Are you aware that DEI has its roots in the Marxist Critical Race Theory?[343]

Do you know that there is no research showing a connection between diversity based on skin color and excellence in military and/or tactical performance? Task Force One Navy had a charge to discover the connection between diversity and excellence in tactical performance and failed to find it.[344]

After the DoD-wide stand-down in 2021, a survey was completed to assess the situation. A senior official told this author, "I reviewed the stand-down brief that was internal to the Defense Human Resources Activity (DHRA), and it included survey results from all of DoD (military and civilian) conducted by the DHRA Office of People Analytics, and the survey indicated that, overall, a scant 2 percent of DoD personnel are concerned about hate crimes or racism." Why are millions of dollars and hundreds of thousands of hours being devoted to implementing DEI?

Do you know that the lag in achievement of some racial groups is not racism but cultural choices, such as a high percentage of single-parent families, having children before getting a high-school education, and not having full-time employment of any type. Data on real reasons for some of our nation's people lagging may be found in the books cited in the bibliography below.

Do you know that research indicates that DEI training programs do not work? Even Harvard[345] says so. And, besides, it's not necessary in our already-diverse Navy.

Do you know that race relations have been improving for decades and that a vast majority of Americans do not feel that race is a major factor in workplace success? The complaint rate for discrimination in the workplace has been going down for decades. See chart on following page.

Is it time for the senior leaders in the Navy, the Admirals, to wake up to the fact that DEI programs mandated by the Executive are unnecessary and harmful to the Navy? There is no evidence the Navy is systemically racist, and pretending that it is harms morale, recruiting, retention, and readiness.

More detailed information on the reduction in racism in America can be found at CM Media in an article titled "Racism in American Is Vanishing: The Facts."[346] *This article appeared in Armed Forces Press.*

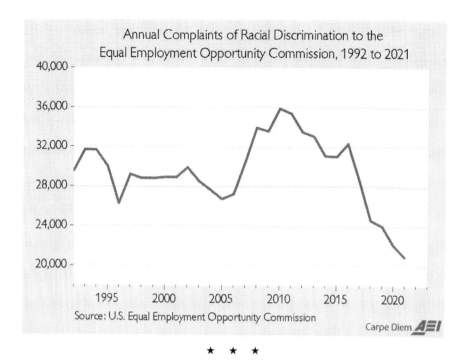

★ ★ ★

The Constitution, the Officer's Oath, Core Values, and Admirals, Armed Forces Press, 24 June 2023

The African Grey Parrot

By Tom Burbage

The Peace Dividend, Lady Liberty's Scales, and the Constitutional Republic Pendulum—words to ponder!

How does a random collection of words, in a language most of us grew up in, have dramatically different interpretations across the panoply of human priorities? According to *Webster*, "language" is the system of words or signs that people use to express thoughts and feelings to each other. So, why this grouping?

African Grey Parrots are the most intelligent and smartest talking birds in the animal species. They attach to only one person. It is up to the owner to teach them because they shape their speaking ability. They have no original thought capacity and can only repeat what they are taught.

The Peace Dividend has historically been referred to as a contraction of the Defense Budget following an extended time of strife and military action. "Peace" becomes a convenient excuse to promote budget priorities to shift from Defense, the primary responsibility of the U.S. government, to other social or domestic priorities. It has taken a dramatic new turn today.

Lady Liberty has always been depicted as being blind to lawbreakers, and her scales have always represented the delicate balance between innocence and guilt. Today, there seems to be a new welcoming of "thumbs on the scale."

The Constitutional Republic Pendulum envisioned by the Founding Fathers created a system of checks and balances that would allow exceptions but provide a self-correcting mechanism to bring the nation back to a central guidance that would sustain the value system of the nation.

So, how do these words help define the challenge our nation faces today? Can they help increase awareness of the real precipice our nation rests precariously on?

The real world follows the fundamental "Laws of Nature" and often succumbs to the "Forces of Evil," but our patriotic DNA trusts that others feel the same priorities we feel. "We trust but don't verify," to steal a theme from President Reagan.

When the deciding factor assumes that our country is defined by the brilliance of the Founding Fathers and the patriots that came before us, our lives and our basic understanding of the "culture of our nation" make perfect sense. When that brilliance is tarnished by the infection of the Woke movement, new upcoming generations reject that premise. We have always had an assumption of trust that our leaders would sustain that culture. That assumption is flawed.

In a way, we are becoming a nation of Grey African Parrots, who can repeat incessantly the mantra of "our trainer." "If we say it often enough, people will eventually believe it is the truth." It is the new reality of trying to interpret the news. When was the last time you heard about the debacle of the withdrawal from Afghanistan or the highly questionable election dynamics? Gaslighting is the new religion of politics. It is a two-phase strategy.

First, don't talk about the bad news; rely on the erosion of facts over time and the short attention span of the American consumer. The combination will eventually make anything a non-factor. Distribute and preach a series of sound bites, and the African Grey Parrots will repeat it to the point of extreme irritability.

Second, talk about the opposite narrative. "The most impressive exit in history" is a recent one—no slight to the great work done by the USAF C-17 fleet. But how many are aware of the high-risk patriot elements that went back into Afghanistan to rescue the Americans left behind? Benghazi and Afghanistan are both great examples of creating narratives to hide reality.

The U.S. Defense Budget priorities are always preeminent in the top-tier responsibilities of the U.S. Congress and the Defense Committees on Capitol Hill. While some believe the U.S. has been in a relatively peaceful time in our history, according to the mainstream media, a new infection has engaged our Constitutional Republic. The "Squad and Far Left Agenda" has introduced another new challenge that was unforeseen by our forefathers. Diverting scarce resources to the introduction of a new Marxist-oriented ideology is the antithesis of our Constitution. Oh, by the way: who said we were at peace?

Perhaps there is no more vulnerable icon in our nation than Lady Liberty's Scales of Justice. She is fighting a clear assault on her ability to "balance those critical scales." How does our system of justice excuse the destruction of cities, police elements, and private businesses, including the latest smash-and-grab and the Wild West-recreated theft of trains delivering internet goods? We need to wake her up so that there is a heavy thumb sneaking under her blindfold, poking her in the eye, and pressuring her scales.

The Constitutional Republic Pendulum is an allegory that says our Founding Fathers built in a visionary mechanism with the separation of powers between the Legislative (Congress), Executive (White House), and Judicial (Supreme Court) constructs. It assumed that the forces of evil could not subvert their intentions. Big money and the control of independent freedoms of expression, coupled with the new censorship dynamic of social media were not on their list of potential "foreign and domestic enemies." If we let them, they will destroy us. They are defining the "domestic enemy" the founders foresaw. Left alone, they will be the death knell of the Constitutional Republic.

Saving our way of life will be dependent on understanding that there are three categories of patriots. First, "the activists," willing to take a stand, often at great personal expense. Second, "the aware but not engaged," who are confident that the pendulum or someone else will step up and pull it back to the center. Third, "the apathetic," who really don't care and don't want to get involved. It's clear that the last two underestimate the severity of the current situation, but they may be the majority. Can that sleepy group be motivated to take a stand?

We must not become a nation of African Grey Parrots; *In a world of rapidly increasing threats, we must prioritize national defense as we face the inevitable economic consequences of unbridled spending.* We must remove, with a vengeance, all thumbs in Lady Liberty's eye and all the thumbs on her scales, and we must remove any corrosion on the Constitutional Republic Pendulum, which seems to be rusting far in advance of the laws of nature.

★ ★ ★

Our Voice Doesn't Matter. Does Our Vote?

By Tom Burbage

MANY YEARS AGO, A NUMBER OF FAULT LINES began covertly penetrating the foundation of our Constitution and our American principles. A major one began with the infiltration of Marxist ideology into our Institutions of higher learning, particularly the Ivy League Universities. As graduates became teachers in lower-level schooling programs, the Marxist virus began the slow infection of our children and changed the traditional patterns of learning. Pride in America, patriotism, and belief in a higher being left the curricula. The pandemic forced a change in the learning dynamic as children participated virtually, and parents could look over their shoulders and see what was being taught. It became clear that trust in the school systems was ill-founded. These were the early indications of a new agenda being forced on America by a dark force, funded by extraordinary resources of a very few and supported by a political agenda that most Americans did not fully understand or embrace.

A second fault line began with the 2011 National Defense Authorization Act (NDAA) that expanded the role of women in the military to include ground combat, including infantry and special forces. Failing to recognize the basic biological differences required to succeed in basic combat began the reduction in standards and the erosion of our military fighting edge. The NDAA also began the movement from meritocracy (best candidate succeeds) to equity (equal outcomes trumps the best candidate) by directing all government agencies to begin the transition to the Diversity, Equity, and Inclusion movement. Ingraining these principles in the military institution is contradictory to the military ethos and requires incentives in the highest levels of leadership to make it a reality. The NDAA ensured that this new criterion became imbedded in the selection criteria for the most senior ranks of the military services requiring Congressional approval.

A third fault line began with the mandated vaccines associated with Covid. The vaccine has attributes that are challenged by religious and general-health concerns. In the military, a significant number (in the thousands) have not agreed to take the vaccine. Many of these individuals are highly trained (USAF F-22 pilots; U.S. Navy SEALS), very fit individuals with very valid concerns about the vaccine. The Supreme Court agreed to hear the case, but the timetable was recently delayed until next year due to animosity following the *Dobbs v Jackson* decision. In the interim, a number of these individuals are being kept in a detention environment, performing menial tasks. Some are reportedly being kept in deplorable conditions now, with no end in sight. Do you see the similarities to "prisoners" still awaiting trials in DC jails for their January 6 actions?

So, how do all these complex faults come together and result in the inevitable earthquake that finally shifts the tectonic plates to the utopia some see in our future? They don't.

We now exist in a world where patriotism has been lost on the next generation. Recruiting has become an unwinnable challenge while we are simultaneously forcing a large number of mid-career, highly trained warriors into unwanted discharges for refusing to take a questionable vaccine. Tens of thousands are no longer in the pipeline, and tens of thousands are being discharged. We are on the path to a very hollow force. Our military remains a national-security requirement and has been dependent on an all-volunteer force for the last two decades. That may have to change, and a draft, now including women, may well take its place. It will be a rude awakening for a generation raised on safe spaces, support animals, and divisive rhetoric.

So, what are we doing?

The Calvert Group initiated an awareness campaign early in 2021. Our 350+ members, distributed over 30-some states, initiated a letter-writing campaign as district constituents for both new veteran Congressional candidates and incumbents. We also contacted groups of State constituents,

Senators, and Governors. To date, we have received very few responses and only a handful of meetings with supporting staff.

It is clear that our voice doesn't matter, at least not yet.
A second passion is related to the fact that we are mostly Naval Academy graduates, some with generations of alumni. It is clear that our alma mater and our sister academies at West Point and Air Force mirror the same concerns. The mandated infusion of Diversity, Equity, and Inclusion assumes the Service Academies are equivalent to the Ivy League universities they try to compete with academically. In the view of most alumni, that is not the case. The Service Academies are chartered to produce combat leaders, not doctors, accountants, lawyers, and professional athletes. While the Service Academies for years were lighthouses developing the leaders of tomorrow, they have become weathervanes in the shifting winds of political social experiments.

The major fallout has been a distinct shift in recent years from being a meritocracy-based culture to using equity-based criteria in critical elements of the Brigade: admissions criteria, selection for leadership positions, and selections for competitive programs. Today's Service Academies have a separate Dignity and Respect chain of command, safe spaces, removal of historical names, and an expansion of affinity groups based on separating Midshipmen and cadets based on race, ethnicity, etc. The quality of our leadership core and the ability to protect our nation from its adversaries have never been more threatened.

It is clear that our voice doesn't matter, at least not yet.
Our national leadership seems consumed with the transition to a new world order, sacrificing our Constitutional principles, despite taking oaths of office saying they would defend it.

It is clear that our voice doesn't matter, at least not yet.
But our votes and the system that they are counted in *better* matter.

★ ★ ★

A Path Forward:
For the Navy and the Nation

By Tom Klocek

"IF THERE IS NO EQUALITY OF OUTCOMES among people born to the same parents and raised under the same roof, why should equality of outcomes be expected—or assumed—when conditions are not nearly so comparable?"[347]

"Can you cite one speck of hard evidence of the benefits of 'diversity' that we have heard gushed about for years? Evidence of its harm can be seen—written in blood—from Iraq to India, from Serbia to Sudan, from Fiji to the Philippines. It is scary how easily so many people can be brainwashed by sheer repetition of a word."[348]

As we tried to point out in the foregoing essays, our beloved Navy, as well as the nation, is in crisis. This should come as no surprise. Recent polls show that an overwhelming majority of Americans feel that the country is moving in the wrong direction—as high as 85% in a poll in June 2022. One daily tracking poll has those thinking the country is moving in the wrong direction currently at 70%, while a Gallup poll in December 2022 has it at 76%.

Another worrisome statistic is that, not only are today's youth not motivated toward military service, but a large number of them are unqualified from both an academic *and* a physical-readiness perspective. Also, there is an increasing lack of trust in government agencies and policies. While a majority of Americans (80%, according to Pew Research) still have confidence in the military, only 25% have favorable opinions of elected officials. Even more telling, Pew reports that trust in government has not exceeded 30% since 2007.

In still another area which relates to recent direction for military personnel, trust in scientists to act in the public interest has also been declining. More importantly, with significant numbers of military personnel released or punished for refusal to take the Covid vaccine, with its rapidly declining effectiveness and its dubious background, this number is even lower for medical scientists (29%). And despite court decisions striking down mandates and the declaration that the pandemic is over, these already-trained personnel are being refused reentry into the service of their country. Significantly, this is happening at a time when the services are struggling to meet recruitment goals, with the Army reducing its goal by 10,000 personnel (25%). In an effort to enlarge the pool of recruiting candidates, some of the services are also modifying their standards (for example, changing body-fat and age requirements).

As concerned citizens and former Naval officers, we have pooled our experience and combined "wisdom" in an effort to find a way forward that would move us away from the precipice of the abyss we are currently facing. While there is no obvious, sure-fire path that will correct the situation, there are several steps that we feel will ameliorate the situation and provide a change of direction for the better.

1. Recognize and accept that there is a problem, for which both sides must work to a common goal.

We must engage with the media and Congress to increase awareness of the decline in our military readiness and strength, and its negative prospects for the future. While many articles exist on a wide variety of media and internet sites, there appears to be no coordinated effort to either come to an agreement on the issue or act in other than partisan ways. We must also engage other retired-military groups to develop a path to solving the problem. The Military Officers Association of America (MOAA) regularly encourages its members to visit their Congressional Representatives. We need to adopt a similar strategy. MOAA currently seems to be avoiding the issue of "wokeness" and its negative impact. We need to engage them on this issue and leverage their resources and experience in influencing policy.

2. Eliminate intolerance as a form of tolerance.

Current training programs to try to eliminate racism actually incorporate a form of racism to fight racism, completely overlooking the old adage that "two wrongs don't make a right."

3. Engage with local government officials, especially school boards and criminal-justice organizations. Remember, "All politics is local."

Many of the problems we are observing have their origin in the education system, some beginning as early as pre-school. Thus, local school boards need to play an important role in addressing these concerns.

4. Take the politics out of science, and return to true scientific method.

Today's science has become too political. As such, it stifles free investigation of the issues and quashes dissent. Rather than collect data in an unbiased fashion, today's researchers focus on data that supports their preconceived conclusions. In some cases, such as environmental areas, data is even manipulated to support the current political position.

Studies that show contrary results are suppressed. For example, the U.S. government, along with teachers' unions and other groups, are continuing their push for encouraging sex-transition chemicals, actions, and surgery for children, despite several countries (Sweden, Finland, and the UK, among others) finding that these actions are of dubious effectiveness and may even be more harmful over the long run, resulting in as much as a 20x incidence of suicide and attempted suicide by the ten-year point. Additionally, science has definitively shown that a person's brain is not fully formed until their mid-twenties. And yet, these "scientists" insist that these children can determine that they were "born in the wrong body." This is a frontal assault on basic biology and encourages the lie that a man can become a woman and vice versa. A basic understanding of chromosomal science debunks such a ridiculous statement, as all of one's cells will contain the chromosomes that identify their birth gender, regardless of what surgeries they undergo.

And, if anyone is uncomfortable with persons with the obvious physical characteristics of the opposite sex, they have no recourse. For example,

Army policy states that completing transition "often does not include surgical treatment." This means it would be considered within regulations for a biological male who identifies as female who has not undergone any sex-change surgery to be allowed to shower with females. A review of the Army training on the subject of gender dysphoria makes one think that a new organization within each Army command will need to be established just to deal with these issues. It is also evident that these policies are a distraction from the commander's responsibility to maintain force readiness.

One way to get away from this is for government support of research to be unbiased. Just like we do double-blind tests of drugs, taste preferences, etc., we should establish bipartisan committees for the purpose of supporting objective research. There are standards within the scientific community for establishing hypotheses and data collection that should be followed rigorously, rather than from a subjective perspective.

5. Remove all policies within the Services that would take actions outside the chain of command.

Most policies that set up diversity, equity, and inclusion (DEI) monitoring include a separate reporting structure that could serve to bypass the chain of command, much like the Soviet Navy had political commissars to report on activities within their ships. As it stands, these DEI chains are, in fact, political in nature and could easily undermine morale as well as incite distrust among crewmembers, thereby being an obstacle to unit cohesion.

6. Eliminate policies that put "equity" (or equal outcomes) over merit in performance, advancement, and assignment.

Most policies of this nature involve or indirectly rely on a quota mentality. By the time someone is of military age, their abilities and attitudes are fairly well developed. It is up to the military to mold that still-malleable individual into someone who can fight on the battlefield. The very term "fight" is a competitive term—someone has to be better than someone else (the enemy). It is a binary situation—someone wins, and someone loses. Yes, over the course of time, there have been wars ending in a draw. However, a review of those cases actually reveals a winner and a loser. Even in the case of a "Pyrrhic victory," the seeming winner at first could actually end up the loser—but there is still a winner and a loser. And, without an available cadre of replacement forces and available equipment, a win against one opponent could well eliminate our ability to compete militarily, at least, on the world stage.

7. Within the military, eliminate ties to political organizations and movements, especially those that undermine the foundational principles of our society (e.g., law and order, the family, unity, etc.) and/or are prejudicial to good order and discipline.

As mentioned previously, the current DEI effort at the Naval Academy features video with a number of Midshipmen touting the Black Lives Matter movement, embodied by the Black Lives Matter Global Initiative. This is a Marxist-based political movement that incites violence and promotes Marxist principles and the destruction of the family. The family is the basic social unit of society and is where most children learn unity and trust, characteristics essential to an effective fighting force.

8. Obtain support from the public through media interviews, articles, public events, etc.

We have a duty to obey the lawful orders of those appointed over us, *but* our allegiance, as indicated by our oath, is to the Constitution of the United States.

Life requires more than a participation trophy. Life is competitive, and to deny that is to deny reality. The whole premise of Darwin's theory of evolution rests on competition, the survival of the fittest. In other words, the fittest who survive merited their survival.

"Education is there to teach you how to function successfully in a complex society, and when you lose sight of that, then you start saying, 'Oh, we've got to concentrate on what kind of pronoun you use, and whether you're really a girl or a boy, and are you a victim?'"—former Housing and Urban Development (HUD) Secretary[349]

★ ★ ★

So, Why Are We Here?

By Tom Burbage

A bunch of angry old men is probably a good description of our inspiration—angry old men who swore an oath to our country and fought to defend and protect it. Many of our classmates made the ultimate sacrifice. At this stage of our lives, we were all ready to fade away and enjoy the fruits of our labors on the golf course, but the fundamental precepts of our lives are now challenged by the reality of an attack on our basic concept of life. We defended it once as young warriors. Our focus now is to make sure the next generation is focused on the same.

It is clear that there is a movement underway to fundamentally weaken the United States as the dominant Democracy and the most powerful military in the world. Most concerning is the relentless assault on the most vulnerable (our K-12 children) and the most powerful (our military) institutions that define who we are. How can anyone deny there is a slow dismantling of both? When both of those are effectively neutralized, the decomposition of the United States and its overarching Constitution will be complete.

The critical question is how far into that transformation are we? What needs to be done to reverse the decomposition, or is it too late? Who believes the critical institutions charged with developing the leaders of the future are immune from this transformation? At what point do parents stand up and demand change? It appears that time is now, but momentum is easily lost if the mountain seems too high to climb.

At what point do military leaders "throw their stars on the table" or ... at what point does the unwillingness to do that become a signal of surrender? It appears we are not there yet, with the exception of a rare few junior officers looking for accountability and finding none. The Afghanistan debacle is the penultimate example, and it is quickly disappearing in the general public field of view. The families of the 13 lost and the 20 or so U.S. military who were grievously injured haven't forgotten, but there is a sense of "Nobody else really cares" in the general public. We have an obligation to not let anyone forget.

Why is there no interest in protecting the rights of patriots who thought they were defending our Constitution and the right to a fair and free election but then found themselves in a gulag-prison situation that denied their Constitutional rights? Where is the Kinte kneeling protest? Why do we allow massive protests for a thug, including an absurd Congressional kneeling in Kinte robes in the Capitol of the United States, commemorating his death but with no recognition of the Afghanistan sacrifice of our military?

What does the current emphasis on "White Rage and White Extremism" and the new focus on "Diversity, Equity, and Inclusion" really mean? The Superintendent Letter to the Brigade in September 2020 indicated a need to stamp out that dimension within the Brigade of Midshipmen. When discussed with the Superintendent a year later, he stated that no evidence had been found within the Brigade. In the interim, an office of DEI and an infrastructure of "Diversity Peer Educators," or DPEs, has been installed. The basic Leadership Pillar of the Naval Academy that flows from the Commandant through the Battalion and Company Commander/Officer now has a separate channel for resolving "DEI Issues." How do these insidious changes on the margins, taken singly as insignificant, when, in fact, taken in the aggregate, support the concept of radical change and create a time-sensitive doubt in the future of our country?

So, why are we concerned, and where do we go from here? It is almost as if the Founding Fathers and everyone who has served and died in defense of our Constitution have suddenly become irrelevant. Despite the fact that all of our leaders take an oath to defend and protect it against foreign and domestic enemies, it is not happening today.

The Obama regime, despite its apparent focus on diversity as a clear signal that we, as a nation, were moving into a new generation of inclusion, took a hard left turn and began the generational drift to the current situation we find our nation in. The slow purge of warrior leaders in the military was the beginning but was not visible or relevant to most of the nation. *Trust but don't verify* was a serious fault in our democratic DNA against domestic assailants despite President Reagan's caution with foreign enemies. We, as a nation, forgot the fact that our Constitution clearly warns of domestic enemies as equally powerful as destructive forces to our future. The domestic enemies have clearly arrived, and we seem to be powerless to rebuke them.

So, what do we do?

1. In the immediate, parents must continue to challenge the School Boards and recapture the educational processes of our children and grandchildren. We cannot afford to lose momentum.

2. In the longer term, there must be an awakening to the reality of the cliff we find ourselves on as a nation. We must make sure that every valid citizen gets a vote, that every vote counts, and that the ability to wrongly influence our election process is fixed once and for all. The 2020 election will someday be a Harvard Business School Case study in the ability of criminal elements to change the outcome of a democratic process.

3. Make the Congressional process defend and protect the leadership-development requirement for the future military, government and industry leaders who come through the Service Academy process. This requires a focus on development of warrior leaders and a return to a meritocracy-based, diversity-sensitive approach.

4. Alumni of institutions that contribute the funding that is the lifeblood of their alma maters need to seriously challenge the future objectives of their universities. Money matters if there is a chance to change priorities.

5. For the Naval Academy initiatives specifically:
    a. Pandemic reality a major challenge
    b. Alumni Association/BGO feedback is telling
    c. CRT is an imbedded cancer, not an academic course

d. Why is there a DEI alternate Chain of Command? What is DPE training? What is the ECA policy? Are these adding value to developing the leadership of tomorrow?
e. Is there equal opportunity for white males for leadership positions?

★ ★ ★

Conclusion

A Call to Action for Patriots

THIS COLLECTION OF ESSAYS was written out of a deep sense of love and devotion for our great country. The authors represent more than 300 years of military service. Some served in combat; all put their lives at risk. Not only is Naval service honorable, it is inherently dangerous in both peace and war. All put their lives on the line to protect the liberties that are all too often taken for granted. For many of the authors, service to the Navy and the nation was and is their life's work. Each of us took an oath to protect and defend the Constitution against all enemies, foreign and domestic. This book of essays was written because we are concerned it is now domestic enemies that represent the greatest threat to our nation. We hope that these essays serve as a clarion call to a sleeping nation that does not yet understand the danger of a weakened Navy.

As we consider the state of the Navy, its eroding readiness, the internal division that is growing, the politicization and distractions that damage cohesion and focus, and the growing recruiting crisis, we recognize more than ever just how important is the oath we took so many years ago. Fundamental to our nation's survival is upholding our Constitution. For that reason, the authors of the essays used the word "Constitution" more than 200 times, and it is central to our thinking. It is the Constitution, more than anything else, that makes our beloved country exceptional. Anyone who argues to the contrary is either ignorant of our actual history or understands it but is loyal to ideas that are contrary and disloyal to our founding values.

We choose to stand up and be counted. We are motivated by the recognition that we are at war with a pernicious ideology, rooted in Marxism, that will destroy our country if not eliminated. It is a war that we did not anticipate and never expected to fight, but fight we will. We affirm that our oath was not voided when we took off our uniforms. It was and is a lifelong commitment. The Navy and Marine Corps core values are "Honor, Courage, Commitment." We honor our vow to protect and defend the Constitution, have the courage to face this existential threat to our Republic, and are committed to the defense of our great nation with every ounce of our strength. We hope that, after reading this volume, you will have the honor, courage, and commitment to join us.

★ ★ ★

A Time of Challenge—Moving Toward the Light

By Tom Klocek and Brent Ramsey

Fr. John Catoir said, "When individuals are responsible, society prospers; when they lose their sense of responsibility, society decays."[350] Also, the late Bishop Fulton J. Sheen said, "We allow the good but tolerate the evil that undermines it and call it 'broad-mindedness.'"[351]

While humanity continues to attempt to exert its own control over creation, with a resultant loss here in America of our traditional and foundational spiritual guidance among many, it has also lost the eternal values proper to the enterprise. Under the guise of concern for all and their dignity, now it endorses and is a proponent of undignified actions, lifestyles, and ideas that fundamentally devalue our humanity and reframe harmful things as good. There is a concerted effort to suppress religion and free speech, and to destroy the family. The Movement for Black Lives (with Black Lives Matter the most prominent among them) publicly advocates destruction of the family and the destruction of capitalism—yet received billions of dollars in support from businesses and individuals believing they were supporting something good. Dr. Martin Luther King, Jr. in his landmark speech of August 28, 1963, said, "I have a dream that my four little children will one day live in a nation where they will not be judged by the color of their skin, but by the content of their character." How far have we traveled from Dr. King's ideal?

This is the world today, and it is a time of encroaching darkness. There are still conflicts occurring throughout the world, some more bloody (Ukraine, Israel) than others (China/Taiwan). At times, the light at the end of the tunnel seems like it's receding from us. But there are points of light—not everyone has succumbed to the darkness.

Voltaire said, "So long as the people do not care to exercise their freedom, those who wish to tyrannize will do so; for tyrants are active and ardent and will devote themselves in the name of any number of gods, religions and otherwise, to put shackles on sleeping men." *"All that is required for evil to triumph is for good men to do nothing"* is variously attributed to Edmund Burke, John Stuart Mill, or Leo Tolstoy and captures what we are up against.

Yes, like all humans, we are flawed. But we continue to look to the light. There are those who would deny God, saying He has no place in the governing of nations. Yet, history has shown that those nations which support religious freedom and whose populace believe in God and rely on Him have done the most for the world. "If you read history, you will find that the Christians who did most for the present world were just those who thought most of the next . . . It is since Christians have largely ceased to think of the other world that they have become so ineffective in this. *Aim at Heaven and you will get earth 'thrown in'; aim at earth and you will get neither.*" (italics added) C.S. Lewis, *Mere Christianity*.

Our national motto is "In God We Trust." God is part of our founding documents as stated in the Declaration of Independence. It has been said that there are no atheists in foxholes. When face to face with one's own mortality, humans often turn to their Creator, the same Creator mentioned in that Declaration. Founder John Adams stated it unequivocally, **"Our Constitution was made only for a moral and religious people. It is wholly inadequate to the government of any other."**[352]

Denying God, denying science in the name of science, totally ignoring the need for responsibility for one's actions in the name of "freedom," and using racism to "fight" racism are not answers. They all contribute to the problem. The Progressive ideology undermines the basic principles that bring this nation and any nation together. Using disunity (divisiveness) as a method to promote unity will not work. Benjamin Franklin reminded the founders, "United we stand, divided we fall." Our nation is one. *E Pluribus Unum*—out of many **one**. This statement is also on our currency and on the Great Seal of the United States. We need to fight against those efforts to divide us. "Every kingdom divided against itself is laid waste, and no city or house divided against itself will stand." (Mt 12:25)

Diversity ideology demands breaking down societal and organizational unity as each individual's racial, gender, or other element of identity becomes the principal determining factor of what is important. When the individual and identity groups become more important than the goal or mission or national culture, cohesion breaks down, and the effectiveness of any organization, any military unit, and society itself is impaired. At the same time, this perverse ideology seems to be leading people more toward Communism and Statism, in which the individual is unimportant beyond as a cog in the machinery of the state.

This collection of essays is an effort to bring that light out of the darkness and make it known to others, so that others who have concerns for the future, the "aware but not engaged" and the "apathetic" will awaken and come to the defense of this great nation. Our nation has blessed the world for more than 200 years with its progress and strength, and, with the blessings of God, wrenched the world out of two world wars and continually aided people suffering from catastrophes all over the world.

We need to awaken the sleeping. We need to inspire our fellow citizens to look to the words of Lincoln's Gettysburg Address and *act*: "**It is for us the living, rather, to be dedicated here to the unfinished work which they who fought here have thus far so nobly advanced. It is rather for us to be here dedicated to the great task remaining before us that from these honored dead we take increased devotion to that cause for which they here gave the last full measure of devotion—that we here highly resolve that these dead shall not have died in vain, that this nation, under God, shall have a new birth of freedom, and that government of the people, by the people, for the people, shall not perish from the earth.**"

★ ★ ★

"A Time of Challenge, Moving Toward the Light" originally published on PatriotPost.US on 30 June 2023 under the title "Dark Times: A Time of Challenge."

The Myth of an Apolitical Military: A Call to Action

By Phillip Keuhlen and Brent Ramsey

MORE THAN A CENTURY AGO, the Progressive leadership of the Democrats seceded philosophically from the Founding Principles of the American experiment.[353] Rejecting the principles of natural law and unalienable individual rights, they adopted the statist relativism of Jeremy Bentham's Utilitarianism and that of his philosophical heirs. They embraced the State as the arbiter of a changeable "common good" and government by the State as preeminent over individuals, regulating personal liberty bestowed by government so long as it supports the state-defined "common good."

Founded upon Woodrow Wilson's paradigm of a "living constitution" with lawmaking by judicial fiat or Executive Order and day-to-day governance by an unelected, unaccountable, administrative state, the proponents of Progressivism have unceasingly advanced a vision and values directly opposed to, and profoundly irreconcilable with, those the country was founded upon. They have mounted a Second American Civil War by a long march through the institutions of governance, education, culture, commerce, and, more recently, science and the military, their Grand Strategy to incrementally subvert the Madisonian Constitution and the Founding Principles it was designed to protect. Their strategy is supported by tactics of lawfare, carefully calibrated civil violence, and institutional subversion.

Americans have a sense that things have gone wrong, but, by and large, their perception is fragmentary, focused on one tactical issue or another. Their focus on individual issues inhibits understanding of how one relates to another and how all support the profound Progressive subversion of American constitutional governance. Shellenberger and Boghossian's "WOKE RELIGION: A Taxonomy"[354] demonstrates the scope and interconnectedness of different fronts in this Civil War for the soul of the American republic. Readers are urged to study the taxonomy, which can be found at footnote ii, below.

In the field of governance, Progressives have mounted sustained legislative, judicial, and executive programs that have the aim of compromising essential elements of Madisonian Constitutional governance, effectively replacing it with an antithetical political philosophy that seeks to divide and conquer. In much of public life and discourse, that philosophy is represented by Critical Race Theory and allied movements and implemented via programs of Diversity, Equity, and Inclusion (DEI). Now, this poisonous ideology has captured our military to the detriment of focus on readiness and lethality. The background and experience of our authors enables the essays in this book to focus on what is happening in and to the military and why it is a danger to our nation. **However, it is important not to lose sight of this battle as part of a greater strategic effort to subvert the Madisonian Constitution and the Founding Values it was designed to protect.**

By functionally choosing sides and thrusting the entire military establishment into a conflict pitting our Founding Values against Progressive ones, the civilian *and* uniformed leadership of DoD have made a fiction of the principle of an apolitical military that has been the basis of civil-military relations throughout the second half of the 20th century.[355] To the extreme detriment of readiness, lethality, retention, morale, and recruitment, the military has been transformed into a political organization with racial quotas. It has degraded ground-combat-unit readiness by lowering standards to allow assignment of women. It has allowed gay-pride celebrations, and transgendered individuals have been recruited and placed in limited-duty status, degrading unit readiness. It has provided transitional surgery while in service and promotion of abortion for service members and family members in contravention of long-standing U.S. law; it has embraced climate change as an existential crisis and top Department of Defense priority. This political fight for the soul of the military has not only devastated morale but also eroded the esteem in which the military has been held since our founding. Trust and confidence in the U.S. military has dropped precipitously, from 70% in 2018 to 46% five years later (-24%), according to a recent Reagan Institute survey. Exacerbating this troubling situation, much of the youth of America no longer wish to serve, having been indoctrinated in false history by our nation's public schools with such revisionist anti-American works as the 1619 Project and Kendi's *How to Be an Anti-Racist*. Even cadets and midshipmen at our Service Academies are exposed to these and similar works as recommended reading. To top it off, the Progressive ideology now in vogue throughout the military has alienated the prime source for recruiting. Young people brought up with traditional patriotic values who are now staying home in droves . . . thus creating a severe recruiting crisis that gets worse each year.

Call to Action:
The authors embrace the viewpoint of President Ronald Reagan,

> "Freedom is never more than one generation away from extinction. We didn't pass it to our children in the bloodstream. It must be fought for, protected, and handed on for them to do the same, or one day we will spend our sunset years telling our children and our children's children what it was once like in the United States where men were free."

Now is the time for all citizens of like mind to stand up, be counted, and fight for the soul of the nation, our liberties, and the preservation of our Constitutional Governance.

Armed Forces Officers must return to the promises they made when they took the Oath of Office to support and defend the Constitution of the United States of America. There is nothing in that oath about diversity, equity, and inclusion. That oath stood us in good stead for more than 200 years.

Citizens, Officers, Retirees, and Veterans must remind military leadership that "just following the civilian leadership's orders" has never been an adequate defense of the indefensible, whether it be DEI as a putative need or benefit with respect to military effectiveness, or subversion of the Constitution and the fundamental American values it protects. EEO is still the law of the land, and it mandates equal treatment. Orders/indoctrination that are counter to that are not lawful orders.[356]

Serving Flag and General Officers must repudiate the politicization of the military, resigning if necessary.

Retired Flag and General Officers must acknowledge the active politicization of the Armed Forces that is underway in support of Constitutional subversion, stop hiding behind the myth of an "apolitical military," and speak out and **LEAD** in defense of the Madisonian Constitution and the Founding American Values it protects.

Medical and legal professionals, particularly in the Armed Forces, must challenge DEI indoctrination as an unethical form of thought reform (i.e., brainwashing)—the application of psychological manipulation without informed consent.[357]

Like-minded citizens must resist Critical Race Theory and its associated DEI programs aimed at destroying American Constitutional governance. ***Fight Back!***[358]

- Understand CRT/DEI
- Challenge CRT/DEI under the law
- Build grassroots resistance movements
- Build broad coalitions
- Get the word out to peers, and especially to legislators
- Engage the churches
- Confront woke institutions
- Stand up to Big Tech censorship
- Monitor federal, state, and local agencies to ensure that taxpayers do not subsidize organizations that are hostile to America and American values
- Develop alternatives to DEI training.

Perhaps it is time for all of us to recall the passages that George Washington had read to the soldiers of the Continental Army at Valley Forge on the eve of their battles at Trenton and Princeton:

"THESE are the times that try men's souls. The summer soldier and the sunshine patriot will, in this crisis, shrink from the service of their country; but he that stands by it now, deserves the love and thanks of man and woman. Tyranny, like hell, is not easily conquered; yet we have this consolation with us, that the harder the conflict, the more glorious the triumph. What we obtain too cheap, we esteem too lightly: it is dearness only that gives everything its value. Heaven knows how to put a proper price upon its goods; and it would be strange indeed if so celestial an article as FREEDOM should not be highly rated.

"I call not upon a few, but upon all: not on this state or that state, but on every state: up and help us; lay your shoulders to the wheel; better have too much force than too little, when so great an object is at stake. Let it be told to the future world, that in the depth of winter, when nothing but hope and virtue could survive, that the city and the country, alarmed at one common danger, came forth to meet and to repulse it."

—Thomas Paine
The American Crisis Number 1
December 19, 1776

This book is about Honor, Courage, and Commitment. We need to restore the concept of **Honor** to our Armed Forces. We need to show **Courage** and call out to the nation and its leaders that we are on the wrong course with the embrace of the false doctrine of DEI. We need a nationwide call upon senior military and political leaders to restore the Armed Forces to its traditional values and eliminate divisive identity politics that weakens us and makes us vulnerable to defeat at the hands of our enemies.

If citizens stay silent, our traditional values of Honor, Courage, and Commitment are undermined, and the Progressive enemies of Constitutional governance win.

It is up to you! ***SILENCE IS CONSENT!***

★ ★ ★

The Myth of an Apolitical Military: A Call to Action, (joint author w/Brent Ramsey), Real Clear Defense, 9 January 2024

Goodbye DEI, Welcome Back MEI

By John Bowen

This final contribution to our book of essays was written in late June, long after all the others. We are now able to see the pendulum beginning to swing away from DEI in the business world as company managements have come under pressure from stakeholders to abandon the senseless push for diversity at the expense of performance and profitability.

Our Armed Forces are another matter. We the People are stakeholders in the military and totally dependent on the Commander-in-Chief to have our best interests at heart with every decision. Those of us who served took an oath to support and defend the Constitution. You saw at the beginning of the book that our efforts are dedicated to the memory of President Abraham Lincoln. His little-known quote bears repeating here and everywhere as it has never been more relevant than it is right now, not even close: ***We the people are the Masters of both the Congress and the Courts, not to overthrow the Constitution, but to overthrow the men who pervert the Constitution***. We in The Calvert Task Group have witnessed this perversion, we are scared for our country, and we are calling it out.

MEI, is Merit, Excellence and Intelligence - three personal attributes historically sought by military recruiters seeking the best candidates to protect and defend our country. Recruit those who have demonstrated **MEI** in their personal lives and all else falls into place.

There are those in today's political arena who have stated their support for fundamentally transforming the United States of America. What better way to accomplish that than to purposely weaken our military by ignoring **MEI** and instead genuflecting before the altar of DEI solely for political gain.

The end result has been predictable: an increasingly weakened military force in leadership, recruiting, readiness and warfighting capability. The bad guys in the world have begun sabre rattling for good reason.

MEI must now serve as the rallying cry that can return our Armed Forces to operating as a **Meritocracy** that has served our country so well for so long.

Final Thoughts

If you have read this far, you should now have a full appreciation for our concerns, and we hope that, like us, you do not take them lightly. One of our authors described us as "angry old men," and we may be out of touch with some aspects of today's world. It is true that most of us have much more time behind us than ahead, but it is also true that we have learned a great deal in our decades of life and that we all, to one degree or another, created this modern world. Societies once revered their elders for their wisdom, and we believe that our lives have brought us some wisdom, too. While our Minority representation is minimal, some of our ancestors also lived disadvantaged lives as serfs, indentured servants, or even slaves. As we have attempted to explain, our family histories and pigmentation are no more important than our height, weight, hair color, or shoe size. None of that matters. Every person deserves to be treated with dignity as befits their humanity and with the respect that they have earned.

With regard to any assessment that we are somehow out of touch, we disagree. While we are not social-media influencers or Instagram experts, we have earned a deep understanding of people—working closely with them for decades, as we would say in the Navy, from the deck plates up. Furthermore, we are in still in touch with the characteristics of people and the factors that made this country great: people with honor, courage, and commitment, and factors like a strong dose of merit. As a group, we represent much more than dedicated military service. Our lives also include hundreds of years more of experience in the fabric of America. We are confident that we understand the kind of leadership it takes to fight and win America's wars and to keep this country a true land of opportunity. We are all committed to passing on that knowledge.

In closing, we would once again remind the reader of what we stated in the Introduction of this book—Ronald Reagan's insight,

> "Freedom is never more than one generation away from extinction. We didn't pass it to our children in the bloodstream. It must be fought for, protected, and handed on for them to do the same, or one day we will spend our sunset years telling our children and our children's children what it was once like in the United States where men were free."

When listening to people demanding "change," it is always vital to ask, "*Qui bono?*" Who benefits? The men who contributed to this book do not. Two hundred forty-seven years ago, a group of remarkable men pledged "... their lives, their fortunes, and their sacred honor..." to gain real freedom for America. We may be in our "sunset years," but our commitment to this nation remains unabated. If America is to lose its freedom, our final thought is and will remain,

NOT ON OUR WATCH.

★ ★ ★

Authors

Randy Arrington:
Cdr. Randy Arrington is a graduate of UCLA, a retired Navy Attack Pilot, and University Professor of Political Science.

My Dad, Senior Chief John Arrington, enlisted in the Navy at the age of 16 in 1945. He became a submarine sailor, serving on 5 different Diesel- and Nuclear-powered subs. He made deployments during the Korean War and the Cuban Missile Crisis and was on the first crew that sailed under the North Pole onboard USS Nautilus.

John Bowen:
Mr. John Bowen is a 1969 Naval Academy graduate. He served eight years active duty, was proud to serve, and enjoyed all of it, including a tour as a helicopter pilot in South Vietnam in 1971-72 with squadron HAL-3. After his military service he entered the financial business sector. He subsequently founded a successful investment management company, as a sole proprietorship for 27 years.

My Father was a Naval Academy graduate, class of 1930. He was a pilot and had the distinction of later being the Commanding Officer of USS Essex CVA 9 when she came out of the yards in 1955 as the Navy's first angle-deck aircraft carrier.

Tom Burbage:
Capt. Tom Burbage is a Naval Academy graduate, former Navy Test pilot, and industry leader, who was responsible for development of the F-22 and F-35 fighter aircraft.

My father was USNA, Class of 1941. He was Officer of the Deck on USS Detroit *at Pearl Harbor on December 7, 1941, when the Japanese attacked. He served aboard the* Detroit *until detached for flight training. While in Fighting Squadron Five, attached to the aircraft carrier* USS Franklin*, he was aboard when she was hit by a kamikaze, resulting in twenty-four hundred killed or wounded. After WWII, he commanded an Air Wing and the carrier* USS Franklin D. Roosevelt.

John Cauthen:

John Cauthen is a U.S. Naval Academy graduate, former naval aviator, and uniformed instructor in the Naval Academy's History Department, teaching Naval History and Western Civilization.

Both grandfathers served in WWII, one as a B-24 pilot for the Army Air Corps in the Pacific Theatre and the other a Chinese Kuomintang officer who first fought against the Japanese and then the Chinese Communists. It was their stories of sacrifice and fighting against tyranny that inspired me in my youth to serve.

Bruce Davey:

LCdr. Bruce Davey is a Naval Academy graduate, former Navy fighter pilot, and former Blue Angel.

Dad was a '39 graduate of USNA, a submariner in both Atlantic and Pacific with eleven war patrols, finishing the war as skipper of Baleo *(SS-285). Awarded both Silver Star and Bronze Star.*

Guy Higgins:

Capt. Guy Higgins is a Naval Academy graduate. He served thirty years on active duty, with five deployments and more than 4,000 flight hours. He worked for Boeing for twelve years as a VP before running his own small consulting business for seven years.

Mike Hollis:

Capt. Mike Hollis is a Naval Academy graduate, former Navy pilot, Naval Aeronautics Engineer, and Vietnam veteran.

My dad enlisted in the Navy during WWII at age 17 and served on the flight deck of the USS Boxer and as an air crewman in torpedo bombers. Ours was a Naval Aviation family, and it seemed totally natural that I would go to the Naval Academy and then to Pensacola to get my wings.

Phillip Keuhlen:
Cdr. Phillip Keuhlen is a Naval Academy graduate and nuclear-industry senior manager. He had the privilege of commanding the USS Sam Houston (SSN-609), a nuclear submarine. He writes on topics related to governance and national security.

I am the son of a son of a sailorman, following proudly in the footsteps of a pioneering WWI aviation CPO and a battle-tested WWII submariner.

Tom Klocek:
Cdr. Tom Klocek is a Naval Academy graduate, former Surface Warfare Officer and Oceanographer, and a former Principal Engineer at a not-for-profit federally funded research-and-development center.

Dale Lawson:
Cdr. Dale Lawson is a Naval Academy graduate, call sign Runt, retired Navy fighter/test pilot, and company president.

I am the proud son of an Army Infantry hero who rose from Private to LtCol in the Army. He was the CO of L Company, Third Battalion, 115th Infantry, 29th Division on D-Day Omaha Beach. During this he was awarded a Silver Star and a Bronze Star, both with combat V, as well as the Purple Heart. I grew up wanting to go to West Point and received a principal appointment to both USMA and USNA.

Bill McCauley:
Mr. Bill McCauley is a graduate of the U.S. Naval Academy and former Navy helicopter pilot. He is a retired partner from the First Boston Corporation and recently retired as CEO of the hedge fund III Capital Management. Bill currently serves as Chairman of Downtowner, an industry leader in micro transit.

My father volunteered for the Navy at age 17 and deployed aboard two heavy cruisers, Philadelphia, and Pittsburgh, in both the Med and the Pacific as a Signalman. Following kamikaze attacks during the battle of Okinawa in March 1945, Pittsburgh was caught in a typhoon and lost approximately 100 feet of her bow.

Bill Newton:
Mr. Bill Newton is a Naval Academy graduate, former Navy pilot, and former Naval Academy Class of 1969 President.

My father was in the Army Air Corps, and then later what they renamed to the Air Force. He flew gooney birds in Korea and received one Distinguished Flying Cross.

Michael Pefley
Col. Michael Pefley is an Air Force Academy graduate, a former Air Force and Air National guard pilot, and industry leader. He was formerly the Executive Vice President for Operations for Stand Together Against Racism and Radicalism in the Services (STARRS). He was also the 1978 Wing Top Gun for the 401st TFW @ Torrejon AB, Spain (F-4 Phantom II).

"I'm the grandson of a French WWI artillery officer who was awarded the "Legion d'Honneur" and "Croix de Guerre" medals.

Brent Ramsey
Capt. Brent Ramsey is a graduate of the University of Nebraska NROTC. His career in the Navy spanned almost 40 years, including command twice and as Executive Director, CBC Gulfport. He is currently an officer with Calvert Task Group, on the Board of Advisors for STARRS and the Center for Military Readiness, and Member and Secretary for the Military Advisory Group for Congressman Chuck Edwards (NC-11).

My father, Brent Ramsey, saw extensive combat aboard the USS Hobson *during WWII, including in the invasion of North Africa, the invasion of Southern France, the invasion of Normandy, and the Battle for Norway. In the Pacific, the* Hobson *fought in the battle of Okinawa, surviving a kamikaze strike while shooting down three kamikaze aircraft. He served during the Korean war and the Vietnam war, retiring as a Master Chief in 1969 after 27 years of service.*

Scott Sturman:

Dr. Scott Sturman is a distinguished graduate of the Air Force Academy, where he majored in aeronautical engineering. After a tour flying helicopters, he attended medical school and spent a career in private practice medicine.

My father served in the U.S. Army infantry in WWII and Korea. His combat experience included campaigns in the Aleutian Islands, Marshall Islands, Leyte Gulf, Okinawa, and Heartbreak Ridge. He was awarded the Silver Star, three Bronze Stars, three Purple Hearts, and the French Croix de Guerre with Palm.

Jim Tulley:

Capt. Jim Tulley is a Naval Academy graduate, former Navy pilot, Naval Engineer, and former Mayor of Titusville, Florida.

My Dad was in the Navy for WWII years only—discharged in 1948 as MM2. I remember him telling of serving aboard USS Ellyson *("Elly Mae"), DD 454, that sank a German U-Boat (U-616) but picked up survivors. After the war, the two crews became good friends and held joint reunions every other year, alternating between Germany and the USA. My father's pride in his Naval service was a major reason for my desire to attend the U.S. Naval Academy.*

Author *Emeritus*
Dick Stratton:

Captain Stratton commenced his military career as a PFC in the Massachusetts National Guard and enlisted in the navy as a Naval Aviation Cadet. He flew combat missions in Vietnam and spent over six years as a Prisoner of War. He was the recipient of the Silver Star, Bronze Star and Purple Heart amongst other awards.

His last six years on active duty were as a part of the USNA family first as Deputy for Operations at the Academy and then as Director, Naval Academy Preparatory School, NETC Newport RI. His second career was as a Licensed Clinical Social Worker specializing in Children and Families and PTSD afflicted combat veterans. He served as President, NamPows Inc. from 1983 to 1985 and Chairman, Veterans Affairs Advisory Committee on

Prisoners of War from 1989 to 1995. He is the father of two combat USNA graduate Marines, the husband of the first DASN(P&F) and a Naval Aviator.

My dad enlisted in the Navy underage in WWI, serving as a Seaman Second on the USS Mercy. He served as a 1st Sergeant of a Massachusetts State Guard Motor Transport Squadron in WWII. He inculcated by word and example into his sons' being that "Stratton men served in the armed forces." He taught that such service was a duty, a privilege, and an honor in repayment for the gift of freedom. Living in the City of Presidents, Quincy MA, the public school system emphasized his stance with lessons of the struggles of the American Revolution. Joining the Navy living under the traffic pattern of NAS Squantum (Shea Field) made my branch choice a no brainer. I have no regrets.

★ ★ ★

Appendix

Open Letter to Naval Academy Alumni Association

Dear Mr. Webb (Jeff)
28 September 2022

Over the last couple of years, there have been ominous storm clouds brewing on the horizon for our nation, the Republic that underpins it, and the institutions that define it. As a community of patriotic Americans, we are faced with a choice: Do we tolerate failed leadership and sit back quietly to await the predictable degradation of the Naval Academy, or do we speak up and call for a bold change of course, one that aligns our Naval Academy on a path that ensures the fleet and Marine Corps are provided top-notch, ethically balanced, and well-rounded officers to lead our nation in times of crisis?

To provide perspective on why change is a mandate in the eyes of many alums, let's look at what has transpired in our nation in a relatively short period of time.

The Covid pandemic awakened many parents to the trust they had misplaced in the hands of their children's teachers. Parents became acutely aware of Marxist-based ideologies being taught in public schools. Vocal rebellions of parents against school boards became common across the country. Sadly, despite the protests by parents, the damage was done. The proliferation of divisive doctrine, open advocation of sexual experimentation to children and advocation of LGBTQ and transgender objectives, coupled with the indoctrination of young students in "CRT" and "DEI," have become a substitute for reading, writing, and arithmetic.

At the service level, the cultural change began in earnest with the National Defense Authorization Act (NDAA) of 2011. This legislation fundamentally altered the nature of our nation and our military, resulting in a new, very different recruiting pool. Today, the societal impact of obesity and the proliferation of drug use, coupled with the erosion of patriotism in our schools has resulted, unsurprisingly, in a sharp reduction in both qualified and interested candidates for military service and, more critically, for the Service Academies.

The unabashed support of this transformation by our current administration, while also pushing out highly trained personnel (like Air Force pilots and U.S. Navy SEALS) for their unwillingness to take a vaccine ensures that we are on a fast path toward a very hollow force, very soon. The recent Congressional push to open military recruiting to HIV-positive personnel adds to these challenges. Training and fielding a military force optimized for combat is not the path we are on.

At the Service Academies, once described as lighthouses for leadership development, there are troubling signs that they may be better described as weathervanes for social engineering.

As loyal alumni of the Naval Academy, we are concerned that the Naval Academy leadership is embracing much of this ideological change and that it is contradictory to the development of combat warriors.

What role does the Alumni Association play in all of this? What role should it play?

When the longtime warrior Tecumseh was renamed, did the Alumni Association express an opinion, or did they give it a pass in order not to ruffle any feathers? When the next group wants to scrub some of the battles on the Naval Academy stadium walls to show only *their* definition of "Just Wars," should we, as alumni, object?

Some alumni are stopping their contributions, both personal and estate, because there are issues at the Academy that do not reflect their values. Some recent graduates do not wear their class rings, once a proud tradition for previous generations of graduates. A growing number of graduates are reluctant to recommend attendance to next-generation applicants. The apparent bias against admission based on merit is a significant factor in all these issues.

While the Naval Academy is subject to policy and orders from the chain of command and the President, the Alumni Association is not. In the opinion of a growing community of alumni, the Alumni Association has a moral obligation to support the greater good of the institution.

The fundamental questions:

1. Should the broader group of alumni be concerned about maintaining the standards required to develop effective leaders of tomorrow?
2. Is the mission of the Naval Academy being faithfully executed?
3. What can the Alumni Association do to positively influence policies and culture at our alma mater?
4. Should the Alumni Association have a larger voice?

But pointing out problems without offering possible solutions is of no value.

***Recommendation:** The Alumni Association should establish a focused subcommittee evaluating the ongoing cultural changes within the institution of the Naval Academy. The subcommittee should review cultural impact on the strategic pillars of the mission of the Naval Academy: Admissions, Academics, Athletics, and Leadership. Representation on this committee should be composed of cognitively diverse, concerned alumni.*

Respectfully,
CAPT Tom Burbage USNR (Ret), USNA '69

★ ★ ★

Follow-up Letter to Naval Academy Alumni Association

Dear Mr. Webb (Jeff)
28 September 2023

One year ago, we communicated and then met at the ASP meeting as you were taking over the challenging position of USNA President and CEO of the Naval Academy Alumni Association and Foundation. Thanks for your leadership during these very difficult times.

The recent article by Rick Hutzell in the *Baltimore Banner* had an interesting comment about older graduates tending to lean right politically. It's unfortunate that every opinion must be wrapped in a political blanket these days. If leaning right means expressing our concerns over the direction our Navy and our nation are going in, then we are definitely leaning right. We do appreciate your commitment that our voices will at least be heard.

I thought it would be appropriate to update you on our concerns and activities, as time seems to help clarify both impressions and concerns.

1. We find that most alumni don't fully appreciate that the Association that most graduates join upon graduation and that many contribute to over their lifetime has no influence over academy programs or policies. As the grandparent of a recent Notre Dame NROTC graduate, I can attest that alumni donations at colleges and universities that have other officer-accession programs do not share this same restriction. The Association's position becomes more problematic when the Board of Visitor charter, which has these responsibilities, can be silenced politically with a change in administration. So . . . should the Naval Academy Alumni Association provide policy input to the Naval Academy beyond fundraising? Certainly, many of the members we have spoken with think we should, even though there may be legal and governmental regulations that govern these dictates.

2. Any challenge to the loss of focus on leadership and unity in the public sphere instantly morphs into an attack on the CRT/DEI movement and the racist label. The question is not whether this Marxist ideology is permeating our school systems. It clearly is, and at *all* the Service Academies, including the Naval Academy. The question is whether the curricula is really teaching about the roots, history, and intent of the movement, or rather endorsing its aims and inculcating its doctrine? If the focus were to teach future leaders to understand why the theory of CRT is not grounded in our Judeo-Christian roots as a nation and is antithetical to the mission of our military, that would be legitimate. Unfortunately, material that is publicly available does not temper the appearance that the Naval Academy and other Service Academies are on a mission of indoctrination, not education. Our concerns cross several areas, including the mission and intent of the Diversity Office and what appears to be a shadow command structure, academic-course catalogs that cover DEI/CRT themes, and the recent hiring post for a professor of Gender and Sexuality Studies. It would be helpful to many alumni to understand the relationship of these excursions to the mission of the premier leadership-development institutions in our military.

3. The Calvert Group recently lost one of our mainstays, RDML Ronald "Rabbit" Christenson, who very unexpectedly left us on August 25th, 2023. Rabbit was the first helicopter pilot to command a nuclear carrier, the *USS Theodore Roosevelt*, and a staunch supporter of the Naval Academy and the Roosevelt Association. In his words, "For the past 15 years, commanding officers have been saddled with flow-down social experimentation from the senior leadership. In addition to—and, in many cases, *instead of*—preparing their unit for combat and sailing in harm's way, commanding officers had to deal with 'Don't ask, don't tell,' introduction of women, zero tolerance, accepting gender dysphoria and trans-sexuality, and, now, identifying extremism, White rage, and Critical Race Theory. It is no wonder that we have so many

COs who have been relieved of their commands. All indicators are that we are proceeding from a meritocracy to 'equity' in our leadership positions. I fear that the brightest will not be there when the Navy and the country needs them the most. As hard as it is to say, I am glad I am not on active duty anymore." For any who knew Rabbit, these are very well-thought-out, careful words. They form the basis of the Calvert Group concerns.

4. Members of The Calvert Group have continued to publish both well-researched articles and opinion pieces in various outlets. A compendium of our articles will be published in an upcoming book titled *Belief Points*, an allusion to the Plebe bible *Reef Points*. The significance of the book is that it incorporates the perspectives of more than a half century of life lessons learned from the impressionable days of memorizing *Reef Points*. Our writers have held leadership positions ranging from Commanding Officers in the fleet, to industry leaders, to members of the bar who have argued cases before the Supreme Court of the United States. They have published dozens of articles raising questions related to the Navy's loss of focus on traditional values and historical leadership principles in favor of advocacy for the DEI movement to the detriment of unity, merit, and recruiting. We hope that the Alumni Association leadership will read and consider the viewpoints presented in the book.

5. Our group of concerned alumni has continued to grow, and we expect to have more than a thousand members by early next year. We have an active website, and, if there is an interest to know more about the Calvert Group, take a look at www.calvertgroup69.org.

6. We remain staunch supporters of the Naval Academy mission to develop the leaders of tomorrow, aligned with supporting and defending our Constitution. Our desire will always be to trust that our institutions like the Naval Academy, the Alumni Association and Foundation, and our Navy have the best interests of our nation and our Constitution at the forefront of their

actions. Joseph Stalin once said that a healthy distrust is a good basis for working together. President Reagan later expanded that thought with the "Trust, but verify" mantra. We live in a very dangerous world, and initiatives at the Naval Academy that erode our warfighting leadership should be a concern to all Americans.

Respectfully,
CAPT Tom Burbage USNR (Ret), USNA '69
President, the Calvert Task Group

★ ★ ★

Further Reading

Know Thy Enemy
    a. Delgado & Stefanic, *Critical Race Theory*
    b. DiAngelo, *White Fragility*
    c. Kendi, *How to Be an Anti-Racist*

Scholars—Alternative Views
    a. McWhorter, "Woke Racism"
    b. Sowell, "Black Rednecks and White Liberals" https://starrs.us
    c. Sowell, "Discrimination and Disparities" https://starrs.us
    d. Sowell, "Social Justice Fallacies"
    e. Sowell, "The Vision of the Anointed"
    f. Steele, "The Content of Our Character"
    g. Steele, "White Guilt"
    h. Swain, "Black Faces, Black Interests"
    i. Swain, "The Adversity of Diversity"
    j. Reilly, "Taboo: Ten Facts (You Can't Talk About)" https://starrs.us
    k. Reilly, "Hate Crime Hoax" https://starrs.us
    l. Hill, "What Do White Americans Owe Black People? Racial Justice in the Age of Post-Oppression" https://starrs.us
    m. Rufo, "America's Cultural Revolution: How the Radical Left Conquered Everything" https://starrs.us
    n. Purdy, "Getting Under the Skin of Diversity: Searching for the Color-Blind Ideal" https://starrs.us
    o. Heriot and Schwarzchild, "A Dubious Expediency: How Race Preferences Damage Higher Education" https://starrs.us
    p. MacDonald, "The Diversity Delusion" https://starrs.us

Resistance
    a. Havel, "The Power of the Powerless"
    b. Lifton, "Thought Reform and the Psychology of Totalism"
    c. Ridgeley, "Brutal Minds"
    d. Swain & Schorr, "Black Eye for America"

Constitutional Governance
    a. Cotton, "Only the Strong"
    b. Goldberg, "Liberal Fascism"
    c. Hamburger, The Administrative Threat"
    d. Magnet, "Clarence Thomas and the Lost Constitution"
    e. Scalia, "On Faith"
    f. Gabbard, "For Love of Country"

Historiography
    a. Gregg, *Reason, Faith, and the Struggle for Western Civilization*
    b. Holland, *DOMINION: The Making of the Western Mind*
    c. Milke, *The 1867 Project: Why Canada Should Be Cherished—Not Cancelled*
    d. Stark, *The Victory of Reason*
    e. Wood, *1620, A Critical Response to the 1619 Project*

Other Authors on DEI/Related Topics
    a. Gonzalez, "BLM"
    b. Lohmeier, "Irresistible Revolution"
    c. MacDonald, "When Race Trumps Merit"

★ ★ ★

Glossary

AOR: Area of Responsibility

BLM: Black Lives Matter

Blue and Gold Officers (BGO): Former officers who act as contact points for young men and women who might be seeking appointments to the Naval Academy

BOV: Board of Visitors

CCP: Chinese Communist Party

CJCS: Chairman of the Joint Chiefs of Staff (Service Chiefs)

CNO: Chief of Naval Operations

CRT/DEI: Critical Race Theory/Diversity, Equity, Inclusion

DHRA: Defense Human Resources Activity

DIE: Variant of DEI; Diversity, Inclusion, Equity

DOD: Department of Defense

DOJ: Department of Justice

DPE: Diversity Peer Educators

Drunk the Kool-Aid: Refers to the "Jonestown Massacre" in 1978, during which a charismatic preacher convinced 900 of his followers to drink cyanide-laced punch, resulting in their death.

EBIT: Earnings Before Interest and Taxes

ECA: Education and Cultural Affairs

GAO: Government Accountability Office

INSURV: Board of Inspection and Survey (Conducts inspections of U.S. Naval ships to determine fitness for service and material readiness to carry out its mission)

LGBT: Lesbian, Gay, Bisexual, Transgender

NGO: Non-Governmental Organization

O'Club: Officers Club (mostly extinct, superseded by Combined Messes)

PLAN: People's Liberation Army Navy
PRC: People's Republic of China (Communist China)
SECDEF: Secretary of Defense
STARRS: Stand Together Against Racism and Radicalism in the Services
STEM: Science, Technology, Engineering, and Mathematics
TF1N: Task Force One Navy
USINDOPACOM: U.S. India/Pacific Command
USNA: U.S. Naval Academy (Annapolis, MD)
USAFA: U.S. Air Force Academy (Colorado Springs, CO)
USMA: U.S. Military Academy (West Point, NY)

Index

Numbers
1619 Project, 16, 70, 307

A
abortion, 17
academic ranking, of Naval Academy, 34
Academy Diversity Reading Room, 119
achievement gaps, cultural choices and, 279
Acton, Lord. *see* Dalberg-Acton, John
Adams, John, 64, 273, 302
administrative state, Wilson's, 50–51
"affinity groups," 24–25, 34, 35, 44–45, 81, 173, 230, 238, 240
affirmative action, 125, 275–276
Afghanistan, withdrawal from, 105, 182, 189, 200, 273, 282, 296
African Grey Parrots, 281
Alexander, Michelle, 60–61
Alien and Sedition Act, 64
alumni
  importance of, 297
  Naval Academy, 33–36, 297
Alumni Association, Naval Academy's, 36, 297, 322–323
  letter to, 321–327
American Federation of Teachers (AFT), 261
amphibious ships, inadequate supply of, 217–218
"anti-essentialism," 250
Antifa, 105, 243
anti-war protests, Vietnam-era, 210
anti-woke activism, 283
Arrington, Randy, 315
Articles of Confederation (1777), 55
Audino, Ernie, 102
Austin, Lloyd, 59, 60, 90, 237, 242

B
Bancroft Hall, 35
Banks, Jim, 112, 120
Becerra, Xavier, 66
Belief Points, 326
Bell, Derrick, 247, 248
Bentham, Jeremy, 49, 52, 305
Berger, David, 217–218
Biden, Hunter, 274
Biden, Joseph, 44, 90, 105–106, 111, 115–117, 203, 234, 242, 278

Bill of Rights, 47
Black Lives Matter Global Initiative, 293
Black Lives Matter movement, xv, 4, 16–17, 86, 100, 105, 119, 137, 205–207, 243, 293, 301
Black Power movement, 247
blacks
  incarceration rates of, 60–61
  in Naval officer corps, 225
  Naval officers, 100, 225
  police killings of, 205–206
Blue and Gold Officers (BGO), 297
Board of Visitors, terminated, 35–36, 89–93, 118, 242, 244
Boghossian, Peter, 306
Bowen, John, 315
Bradley, Scott, 56
"brainwashing," 260–261. see also thought control, DEI and CRT as forms of
Braithwaite, Kenneth, 216
Brown, Anthony, 241
Brown v. Board of Education, 220, 253
Buck, Sean, 238–239, 243
Buell, Augustus C., 2, 161
Bui, Major, xviii
Burbage, Tom, 37, 315–316
Burke, Edmund, 302
Butler, Paul, 249

C

Caesar, Gaius Julius, 63–64
Calvert, James, xv, 237
Calvert Task Group, xvii, 39, 81, 218, 235, 237, 242, 286–287, 311, 326
capability, of Navy, 164–165
capitalism, attacks on, 261, 301
Carbado, Devon, 249
Castro, Fidel, 66
Catoir, John, 301
Cauthen, John, 316
Center for Military Readiness, 218, 235
Center for Teaching and Learning, USNA, 77
Challenger explosion (1986), 12
Chang, Gordon, 233
Chavez, Cesar, 247
Chavez, Hugo, 65
Chicano movement, 247
China
  infiltration of higher education, 173
  spy balloons from, 182
  threat from, 35, 121, 163–164, 190, 192, 213, 233–235
Chinese Communist Party, 233–235
choosing sides, 269–271
Christenson, Ronald, 325
Christians, 81, 120, 123
Churchill, Winston, 70
Civil Rights Act of 1964, 41, 116, 196, 209, 253
Civil War, US, 28–32, 64, 116
civilian leadership of military, principle of, 307
class, as determinate of success, 61–62
climate change, 16, 215

Coates, Ta-Nehisi, 113, 117, 120
Code of Conduct, Naval Academy, 4
"color blindness," 19–21
Columbia-class submarines, 218
combat capability, degraded by DEI, 199–200
commitment, core value of, 278
Confederacy, treason allegation levelled at, 27–32
Constitution
   Article I, Section 8, Clause 14, 91
   attacks on values of, 256–257
   Bill of Rights, 47
   Equal Protection Clause of, 41
   Fifteenth Amendment, 116
   Fourteenth Amendment, 41, 116, 227, 271
   guarantee of national defense, 230
   Madisonian vs. "living constitution," 47–53, 305–306
   practical success of principles, 66
   right to secession, 27–28
   upheld by morality, 169–171
Constitutional Convention, 27, 47–48
Constitutional Republic
   America as, 7–9, 64, 66
   attacks on values of, 283
   checks and balances in, 281, 283
   generational responsibility to steward, 255–256
core values, of Navy, 277–278
Cotton, Tom, 112, 120, 237
Council for Academic Freedom, Harvard's, 140

Counter Extremism Working Group, 111
courage, core value of, 278
Crenshaw, Dan, 112, 120
Crenshaw, Kimberlé, 131, 248
critical legal studies, 247
critical pedagogy, 77
Critical Race Theory, 159, 246
Critical Race Theory (CRT)
   activist component of, 249–250
   in Air Force, 119
   in Air Force Academy, 119
   in Army, 119–120
   basic tenets of, 250
   defined, 246–247
   deleterious effects on military, 200
   early origins of, 247
   education component of, 249
   foundation for DEI, 279
   founding American values and, 306
   goals of, 116–117
   in higher education, 206–207. *see also* particular services academies under this entry
   indoctrination of public school system, 43–45
   insidious nature of, 96, 205–207
   intellectual antecedents of, 247–248
   major themes of, 250–251, 253–254
   merit versus, 70
   in military, 61, 73–75, 106
   in military generally, 234. *see also* particular services academies

under this entry and particular branches of military
military readiness and, 306
national decline and, 56
at Naval Academy, 59, 77–78, 237–244
in Naval Academy, 297, 325–326
in Navy, 112, 190
origins of, 115–116
pervasiveness of, 4
political program of, 251–252
in public education, 322
rejecting values of, 160
resistance to, 308
in Space Force, 119
in TF1N, 145
as thought control, 259–265
undermines equality, 73–74
at West Point, 113, 119–120
Critical Race Training, at Naval Academy, 262
CriticalRace.org, 238, 239
Cummings, Andre, 249

D

Dalberg-Acton, John, 63
Davey, Bruce, 316
Declaration of Independence, 47–48, 116, 160, 230
Defense budget priorities, 282–283
"Defund the Police" movement, 256
DEI Master Plan, Naval Academy's, 196
Del Toro, Carlos, 216
Delgado, Richard, 78, 159–160, 246, 247, 248
democracy, Constitutional Republic vs., 7–9
Democratic Party, rejection of founding American principles by, 305
demographics
of Navy, 107, 122, 211, 224–225
TF1N analysis of, 152–158
deployments, length of, 190
Derrida, Jacques, 247
desegregation of military, 73, 209, 220, 253
DiAngelo, Robin, 240
diversity
diversity initiatives of Navy and, 158–159
as false value, 73–74, 289
merit vs., 219–220
in military, 61
misplaced emphasis on, 185–187
in Naval Academy, 82, 241
in Navy, 83, 97–103, 107, 109, 122, 224–225, 245–246
in professional sports, 220–221
real vs. perceived value of, 146–148
in service academies, 219–222
Diversity, Equity, and Inclusion (DEI)
in Air Force, 112–113, 118–119
in Air Force Academy, 118–119, 130–131, 263, 287
deleterious effects on military, 120–127, 195–198, 200
as disincentive to work, 275–276

founding American values and, 306
goals of, 116–117
in higher education, 140, 142. *see also* particular services academies under this entry
identity-based studies of service academies and, 129–133
impact on Naval officers, 278–279
indoctrination in military and, 56, 245
at Merchant Marine Academy, 131
merit versus, xv, 70, 285–286
in military, 73, 106, 182, 234, 259–265
military professionalism and, 139–143
military readiness and, 112, 130, 199–200, 221, 306
at Naval Academy, 131, 287, 293, 325–326
in Navy, 107–110, 111–114, 118, 168, 174, 190, 195–198, 222–227
no justification for in Navy, 203–204
offices, 74, 82
origins of, 115–116
pervasiveness of, 4, 41–42
political aims of, 252, 254
in public education, 322
reality of threat from, 117–120
replacing Duty, Honor, Country, 35
resistance to, 308
roots in CRT, 279
separate chains of command for, 45, 113, 120, 197, 238, 241, 292, 296, 298, 325

in service academies, 261–265, 287. *see also* particular services academies under this entry
in Space Force, 112–113, 118–119
Task Force 1 Navy and, 59
as thought control, 259–265
undermining trust, 161–162
unity and cohesion and, 79–84
at West Point, 122, 130–131, 287
Diversity, Equity, and Inclusion minor, at Air Force Academy, 130, 221
Diversity, Equity, and Inclusion offices, 221
Diversity and Inclusion Strategic Plan, Naval Academy's, 239
diversity conference, Naval Academy's annual, 81
Diversity Peer Education, 45, 174, 241, 296, 298
"diversity peer educators," 296
diversity peer educators, 241
diversity policy statement, Navy's, 98
Dobbs v. Jackson, 286
Donnelly, Elaine, 214
Douglas, Frederick, 247
Du Bois, W. E. B., 247
Duke University Medical School, 221
Dying Citizen, The, 220, 222, 240

E

"economic democracy," 252
economic equality, 178
education
   indoctrination in, 70

workforce participation and, 62
Education and Cultural Affairs (ECA) policy, 298
election fraud, 2020 presidential election and, 274, 282
election integrity, 174–175, 297
elite athletes, admissions preferences for, 99
Emancipation Proclamation, 178
"empathetic fallacy," 251
Equal Protection Clause, 41, 227
Equal Rights Act (1964), 16
equality
  CRT as undermining, 73–74
  equity vs., 225–226
equality of opportunity, xv–xvi, 229, 275
"equality of outcomes," 130–131, 221, 225–226, 275
equity
  equality of opportunity versus, xv–xvi, 275
  merit vs., 292–293
  in military, 61
  Navy emphasis on, 107–108, 225–226
ethnic studies, 248, 249
Evangelista, Paul F., 102
Evans, Ernest E., 79–80
Executive Order 9981, 220, 253
Executive Order 13583, 115–116, 261
Executive Order 13983, 158–159
Executive Order 13985, 115
"extremism," in military, 59–60, 101, 111

F

family, attacks on, 86, 293, 301
Farragut, David, 1
feminism, 247, 248
Fifteenth Amendment, 116
flag officers, 109, 253, 267
Fleet Admirals, 37–40
Floyd, George, 100, 214
Ford class carriers, 166
formation of sailors, 23–25
Foucault, Michel, 247
Founding Fathers, 15, 48–49, 64, 66, 70, 174, 178, 179, 230, 257, 273, 296
Fourteenth Amendment, 41, 116, 227, 271
Franklin, Benjamin, 52, 302
Freedom of Information Requests, 119, 122, 123
freedom of religion, Progressive attacks on, 51–52
freedom of speech, Progressive attacks on, 51
Freeman, Alan, 247, 248
Fromm, Erich, 49
Fryer, Roland, 125

G

Gallagher, Mike, 112, 120
Garfinkle, Herbert, 142
gender
  diversity initiatives and, 101
  leadership selection based on, 213
Gender and Sexuality Studies, at Naval Academy, 81

George Floyd riots, 214, 243, 296
Gettysburg Address, 303
Gilday, Michael, 100, 165, 192, 223
Gillibrand, Kirsten, 241
Gomes, Laura, 248
Gorshkov, Sergey, 245
Gotanda, Neil, 248
government, function of, 178
Grabar, Mary, 16
Gramsci, Antonio, 247
Grant, Ulysses S., 30
Great Depression, 51
Great Society, 86, 275
Green, Mark, 90
"grievance studies" hoax (2017-2018), 129–130
Guevara, Ernest "Che," 66
Guinier, Lani, 249
Gulati, Mitu, 248
gun violence, 16–17

H

Halsey, William, 37, 39–40
Hamilton, Alexander, 64
Hanson, Victor David, 220, 222, 240
hard word, value of, 178
Harris, Angela, 248
Harris, Cheryl, 248
Harvard University, 221
Harvard v. SFFA, 41, 83, 99, 101, 107
Hegel, Georg Wilhelm, 49, 52
Hegelian state, 50
Henry, Patrick, 53, 205
Herbert, Frank, 63
Higgins, Guy, 316

Hill, Jason, 137
Hispanics, in Navy, 224–225
Historically Black Colleges (HBCs), 221–222
history, falsification of American, 15–17
Hitler, Adolph, 65, 66
Ho Chi Minh, 65
Hollis, Mike, 316–317
Homeland Security, Department of, 56
honor
 core value of, 277, 278, 309
 truth and, 161
Hopkins, Esek, 1
How to Be an Anti-Racist, 112, 118, 213, 307
Hueman, Pat, 122
Hunter, Eduard, 260
Huntington, Samuel P., 140–141
Hussein, Saddam, 12
Huston-Tillotson University, 221
Hutzell, Rick, 324

I

identity-based studies, at service academies, 129–133
illegal immigration, 7–9, 34–35, 56, 122, 182, 256
"implicit bias," 239
implicit bias training, 74
incarceration rates, blacks and, 60–61
inclusion
 Navy emphasis on, 226
 Nay emphasis on, 108–109

Index of Military Readiness, Heritage Foundation's, 234
"interest convergence," 250
interracial marriage, widespread support for, 135, 212
"intersectionality," 249, 250
Iran, 34, 182
Israel, moral decline of ancient, 170–171
Israel-Palestine conflict, 190

J

January 6th protesters, 296
Jefferson, Thomas, 64, 73, 273
Johnson, Kevin, 248
Johnson, Lyndon, 86, 275
Jones, John Paul, 1–2, 161, 195, 230
justice system, Leftist attacks on, 283

K

Kang, Jerry, 248
Keane, Jack, 90
Kendi, Ibram X, 60, 100, 112, 118, 211, 213, 307
Kennedy, Robert, 19
Keuhlen, Phil, 102, 225, 317
King, Ernest, 37, 38, 40
King, Martin Luther, 19, 221, 247, 301
Klocek, Tom, 317

L

Lady Liberty, 281
Lawrence, Charles, 248
Lawrence, James, 1, 230
Lawson, Dale, 317

leadership
 America's need for, 177–179
 in Navy, 165–168, 191–192, 215–216
 in service academies, 297
Leahy, William, 37–38
Lee, Robert E., 28, 30–31
legal indeterminacy, 247–248
legal professionals, military, 308
Lenin, Vladimir, 65
Levit, Nancy, 249
Lewis, C.S., 302
LGBTQI+, in military, 80–81
liberalism, CRT critique of, 251
Lifton, Robert Jay, 260, 262
Lincoln, Abraham, 30, 39, 52, 64, 70, 178, 303, 311
Littoral Combat Ships, 166, 217
"lived experience," 61
"living constitution," 47–53, 271, 305–605
local government, focus on, 291
Locke, John, 52, 178
Lohmeier, Matthew, 113, 119
Lopez, Gerald, 248
Lopez, Ian Haney, 248
Loury, Glenn, 124, 125, 137
Luce, Stephen B., 132

M

Madison, James, 27–28, 48, 64, 177
Madisonian vs. "living" constitution, 47–53, 305–306
Madras, Bertha, 140
Manchurian Candidate, The, 260
Mao, 65, 66

Marcuse, Herbert, 49, 262
Marx, Karl, 49, 55
Marxist ideology, indoctrination with, 8, 16, 41, 73, 86, 113, 116, 119, 238–239, 256, 259, 261, 264, 270, 283, 285, 300, 321–322, 325
Matsuda, Mari, 248
Maury, Matthew Fountaine, 214–215
Maury Hall, renaming of, 214
McCain, John, xviii
McCauley, Bill, 317–318
McKinsey & Company report, 149–152
McMaster, H. R., 44, 91, 92, 233
McWhorter, John, 124
medical personnel, military, 308
Memorial Day, 20–21
merit, diversity vs., 219–220
Merit, Excellence, and Intelligence (MEI), 311–312
meritocracy, 70, 124, 211, 285–286, 312
military bases, movement to rename, 27–32
Military Officers Association of America (MOAA), 290
military professionalism, 139–143
Mill, John Stuart, 49, 302
Minihan, Mike, 234
minorities, privileging of, 45, 99, 100
MIT, 221
Montoya, Margaret, 248
morale, falling, 123
morality, 169–171, 183
Morrison v. Olson, 93

Movement for Black Lives, 301. *see also* Black Lives Matter movement
Mussolini, Benito, 66

N

Napoleon, 65, 127
National Defense Authorization Act, 2011, 285, 322
National Defense Authorization Act, 2018, 192
National Education Association (NEA), 261
national identity, American, 29–30
national security, prioritizing, 170
national service program, 275–276
National Vote Interstate Compact, 51
natural law, 48–49, 52, 305
natural rights, 48, 52, 160
Naval Academy, United States
  formation of officers at, 1–2
  mission of, xviii
New Jim Crow, The, 118
Newton, Bill, 318
Nimitz, Chester, 37, 38–39
non sibi, xvii, 5
Norman, Ralph, 90
Northern Ireland, 19–20
Nowell Jr., John, 103
"nuisance groups," 96
Nuremberg Code, 123

O

oath to Constitution, 2, 7–9, 20, 33, 47–48, 53, 55–57, 74, 77–78, 140,

196, 203, 227, 256, 271, 277, 293, 295, 296, 299, 300, 307
Obama, Barack, 41, 44, 65, 115–117, 174, 203, 213, 234, 261, 278, 296–297
Office of Diversity, Equity, and Inclusion Awards, Naval Academy's, 240
Office of Diversity, Equity, and Inclusion, Naval Academy's, 219, 220, 238, 240–241, 244
officers
  black Naval officers, 225
  blue and gold officers, 297
  criteria for selection of, 83
  DEI programs and, 278–279
  flag officers, 109, 253, 267
  leadership qualities of Naval officers, 166
  Naval demographics and, 154–155
  need to resist progressive agenda, 307–308
  not benefitting from DEI, 109
  oath to Constitution and, 307
  preference for Black, 100
  purged for opposition to DEI, 117
Onwachi-Willig, Angela, 249
Orwell, George, 137

P

Page, Richard Lucian, 31
Paine, Thomas, 309
pandemic, COVID, 85–87, 161, 166, 167, 182, 215, 235, 275, 285, 290, 297, 321

patriotism, loss of, 182
"peace dividend," 281
Pearl Harbor, attack on, 235
Pefley, Michael, 318
Perea, Juan, 248
Peron, Juan, 66
Pierceson, Jason, 61
Pillsbury, Michael, 233
Pinker, Steven, 140
Pol Pot, 65
police killings, allegedly targeted at blacks, 137
political conditions, present US, 273–274
political parties, 21
power, necessity to limit, 63–67
Prelogar, Elizabeth B., 99
Prep School, Naval Academy, 39
Pride Month, 80–81, 101, 190
Prince, Bill, 102
Princeton University, 221
priorities, Navy's misplaced, 213–218
Proceedings of the National Academy of Sciences, 147–148
professional culture, at Air Force Academy, 140
professional sports, diversity in, 220–221
Progressivism, 49–51, 305–306
Province, Michael, 231
public education
  attacks on Constitution in, 56
  decline of, 173
  indoctrination in, 85–87, 199, 307, 321

restoring integrity of, 174
public support, gaining, 293
Putnam, Robert, 61–62

Q

"Qualifications of a Naval Officer," 1–2, 161, 195

R

"race hustlers," overrepresented in the military, 137
race relations, in Navy, 210–212
"race remedies law," 251
racial disparities, actual reasons for, 124–127
"racial gap," closing, 61–62
racial preferences
   in Naval Academy admissions, 82–83
   in Navy, 107
racism
   actual prevalence of, 124–125
   alleged pervasiveness of, 4, 33, 44, 101–102, 111–112, 124, 135–138, 189–190, 211–212, 279
   as foundational to America, 16
   in military, 120, 229
   Navy and, 108
   promoted by anti-racism programs, 291
   West Point and, 132
Ramsey, Brent, 318
Rangers, lowering of standards, 44
readiness, military, 112, 119–200, 165, 215, 221, 230, 235, 290, 306

Reagan, Ronald, 103, 181, 200–201, 282, 297, 307, 314, 327
Reardon, Sean, 62
reconciliation, 181
recruitment numbers, falling, 81, 123, 139, 167, 182, 189–190, 214, 235, 236–264, 270–271, 286, 290, 322
Reef Points, 1, 3–5, 326
regulatory state, FDR's, 51
Reilly, Wilfred, 124, 137
responsibility, generational, 255–258
Restore Liberty, 218, 235
retention, falling, 123, 235
reversing national and military decline, 267–268
"revisionist history," 250
Rielly, Wilfred, 61
Roberts, John, 83
Rogers, Mike, 59–60
Roosevelt, Franklin, 38, 51, 64
Roosevelt Association, 325
Ross, Tom, 249
Rubio, Marco, 234
Rufo, Christopher, 61, 116
Russell, Catherine M., 92

S

Sada, Georges, 12
Sailor's Creed, 203, 277
school boards
   parental revolt against indoctrination and, 85–87
   taking back, 174, 291, 297
science, politicization of, 291
secession, states' right to, 27–28

Second Virginia Convention (1775), 205
Seila Law LLC v. Consumer Financial Protection Bureau, 93
selection criteria, at Naval Academy, 82–83
self-sacrifice, military value of, 230
September 11th, 2011, attacks, 269–270
service academies, restoring integrity of, 175
"serving two masters," 251
sexual assault
  in military, 235
  in Navy, 215
Sexual Minorities and Politics, 118
Sheen, Fulton J., 301
Shellenberger, Michael, 306
Sherman, William Tecumseh, 30, 322
shipbuilding plan, congressionally mandated, 217
shipmates
  formation of, 23–25
  values of DEI vs., 83–84, 240
slavery, in America, 15–16, 29–30
"social constructionism," 250
social engineering, in military, 43–44
Social Equality--NL450 course, Naval Academy's, 239
"social justice," corrosive effects on military, 111–114
socialism, 16, 20, 55, 70, 178–179, 189, 256
Solomon, King, 170
Sowell, Thomas, 61, 63, 124, 125, 137

Spicer, Sean, 90
spiritual values, decline in, 301–303
Stalin, Joseph, 66, 327
Stanford University, 221
STARRS, 79, 112–113, 117–118, 119, 123, 218
states
  national identity and, 29–30
  right to secession, 27–28
Steele, Shelby, 61, 124, 137
Stefanic, Jean, 249
Stirrup, Heidi, 90, 92–93
Stoltz, Merton P., 125
Strategic Imperative One, Naval Academy's, 241
Stratton, Dick, 319–320
structural determinism, 251
Students for a Democratic Society, 210
Students for Fair Admission, 82–83
Sturman, Scott, 319
suicide rate
  military's, 235
  Navy's, 167, 191, 215, 235
Supreme Court, 51
Swain, Carol, 124, 137
"systemic bias," 131
"systemic racism," 101, 119, 124–127, 145, 146, 153, 243, 245, 250

T

Task Force 1 Navy, 59, 98, 100–101, 102, 118, 121–122, 145–160, 204, 225, 245–246, 252, 279
"tenure protection," 91

Themistocles, 63
thought control, DEI and CRT as forms of, 259–265
Thought Reform and Psychology of Totalism, 260
time, perspective on, 69–71
Title VI, Civil Rights Act, 41, 253
Tolstoy, Leo, 302
transgenderism, 17, 44, 51, 59–60, 62, 101–102, 123, 190, 215, 234, 291–292
"treason" allegation, levelled at Confederacy, 27–32
"tribe," military community as, 86
"Troubles," 19
Truman, Harry, 73, 209, 220, 253
Trump, Donald, 66, 90, 115, 117, 242
trust
  institutions and, 289–290
  integrity of Constitutional republic and, 256
  leaders and, 282
  in military, 306
  undermined by DEI, 161–162
truth, virtue of, 12–17
Truth, Sojourner, 247
Tulley, Jim, 319
Turtora, Anthony, 230
tyranny of the minority, 179

U

Ukraine War, 34, 70–71, 182, 190
UNC v. SFFA, 83, 99, 101, 107
Understanding left-wing authoritarianism, 65–66
Uniform Code of Military Justice, 25
unit cohesiveness, 221
United States, Leftist movement to weaken, 295
United States vs. Perkins, 93
"unity ethos," 34, 79–84, 109, 226
US-China Economic and Security Review Commission, 163, 233–234
USNA-at-Large, 218
USS Bonhomme Richard incident, 121, 166, 191, 200, 216–217
USS Connecticut, 121, 191
USS Fitzgerald. see USS McCain-USS Fitzgerald collision
USS Ford, 190, 191, 218
USS Gettysburg, 191–192
USS Johnston, 80
USS McCain-USS Fitzgerald collision, 24, 166, 192, 216, 230
USS Richard L. Page, 31, 210
USS Roosevelt, 166
USS Saratoga, 39
USS Theodore Roosevelt, 325
utilitarianism, 49, 305

V

vaccine mandates, COVID, 106, 113, 122, 161, 167, 264, 286, 290, 322
Valdes, Francisco, 248
Veterans Administration, 21
Veterans for Fairness and Merit, 218, 235
Vietnam War, 11, 210
Virginia Convention (1775), 53

"virtue signaling," 63
Voltaire, 302
Voting Rights Act of 1965, 116

W

Washington, George, 52, 64, 308–309
wealth inequality, myth of, 17
Webb, Jim, 44
West, Bing, 102–103, 217
"white extremism," 296
White Fragility, 74, 240
white liberals, 31
white males, in leadership positions, 298
"white privilege," 73–74, 119, 126, 132
"white rage," 113, 119, 296
"white supremacy," 4, 8, 44, 59, 101, 111, 123, 124, 126, 189–190, 240
Wildman, Stephanie, 249
Williams, Patricia, 248
Williams, Robert, 248
Williams, Walter E., 27
Wilson, Woodrow, 50–51, 305
WOKE RELIGION: A Taxonomy, 306
"wokeism," 44, 77, 229–231, 275–276. see also Critical Race Theory (CRT); Diversity, Equity, and Inclusion (DEI)
women
 in combat roles, 44, 285–286
 underrepresented in Navy, 107
Woods, Peter, 16
Woodson, Robert, 61, 124, 137
Wuhan, China; origin of COVID in, 235

X

Xi Jinping, 233

Y

Yale University, 221
Yamamoto, Eric, 248

Z

Zumwalt class ships, 166, 191, 217

Notes

Introduction
1. https://download.militaryonesource.mil/12038/MOS/Reports/2021-demographics-report.pdf

In Search of Truth
2. Bill of Rights Institute, Slavery, and the Constitution, accessed September 12, 2023
3. "The 1619 Project Gets Schooled," Elliott Kaufman, *Wall Street Journal*, December 16, 2019.
4. Data on Mass Murder by the Government in the 20th century, Dave Kopel, *The Volokh Conspiracy*, November 9, 2022.
5. "Unsettled: What Science Tells Us, What It Doesn't, Why It Matters," Steven Koonin, April 27, 2021.
6. "Trends and Patterns in Firearm Violence 1993-2018," Department of Justice, Bureau of Justice Statistics, Grace Kena and Jennifer L. Truman, Ph.D., April 2022.
7. "Gun Violence Deaths: How the U.S. Compares with the Rest of the World," NPR, Nurith Azeman, January 24, 2023; "How Dangerous Is Lightning?" National Weather Service, accessed September 12, 2023.
8. "Two Million Stopped While Illegally Entering the U.S. from Mexico in 2021," Anna Giaritelli, *Washington Examiner*, January 24, 2022.
9. "Summary of the Latest Income Tax Data, 2021," Tax Foundation, February 3, 2021.
10. "The Transgender Craze Is Creating Thousands of Young Victims," *The American Conservative*, July 21, 2020.

11 https://www.americanthinker.com/articles/2022/09/shipmates_a_dying_breed.html
12 https://www.americanthinker.com/articles/2023/02/truth_in_government.html

Shipmates—A Dying Breed
13 https://www.americanthinker.com/articles/2022/09/shipmates_a_dying_breed.html

Renaming Military Bases Ignores History
14 Treaty of Paris (1783) | National Archives, Article First
15 Bill of Rights Institute, Slavery, and the Constitution, accessed September 12, 2023.
16 S.J. Resolution 23-Public Law 94-67 passed August 5, 1975, A Joint Resolution to restore posthumously full rights of citizenship to General R. E. Lee.
17 Ann Brown, Duke University, August 10, 2020
18 www.richmond.edu, "Slave Trade to the Americas 1650-1860," November 11, 2014.
19 *World Population Review*, 2021.
20 *USS Richard L. Page* DEG FFG 5 Guided Missile Frigate Brigadier General Richard Lucian Page (seaforces.org)

Where Are the Halseys of Yesteryear?
21 Walter R. Bornem, *The Admirals: Nimitz, Halsey, Leahy, and King—The Five-Star Admirals Who Won the War at Sea*, (New York, Little, Brown and Company, 2012)

The View from 25,000 Feet
22 LtG (Ret) H.R. McMaster, "Preserving the Warrior Ethos," *National Review*, October 28, 2021

What Are DOD's Priorities?
23 Michael, Lee, "Pentagon Report Finds About 100 Troops Involved in Extremist Activities," Fox News, December 20, 2021.

24 Geoff Ziesulewicz, "CNO to Stand Up Navy Task Force on Race and Inclusion," *Navy Times*, June 30, 2020.
25 Ryan Morgan, "U.S. Navy Cuts 'Woke' Anti-American Books from Official Reading List," *American Military News*, May 10, 2022.
26 Garance Franke-Ruta, "Robert Putnam: Class Now Trumps Race as the Great Divide in America," *The Atlantic*, June 30, 2012, https://www.theatlantic.com/politics/archive/2012/06/robert-putnam-class-now-trumps-race-as-the-great-divide-in-america/259256/
27 Table 501.10, "Labor Force Participation, Employment, and Unemployment of Persons 25 to 64 Years Old, by Sex, Race/Ethnicity, Age Group, and Educational Attainment: 2016, 2017, and 2018," in U.S. Department of Education, Institute of Education Sciences, National Center for Education Statistics, *Digest of Education Statistics*, 2019, https://nces.ed.gov/programs/digest/d19/tables/dt19_501.10.asp
28 Sean F. Reardon, "The Widening Academic Achievement Gap Between the Rich and the Poor," in *Inequality in the 21st Century: A Reader*, ed. David B. Grusky and Jasmine Hill (New York: Routledge, 2018), pp. 177–189.
29 Ron Haskins, "Three Simple Rules Poor Teens Should Follow to Join the Middle Class," Brookings Institution, March 13, 2013, https://www.brookings.edu/opinions/three-simple-rules-poor-teens-should-follow-to-join-the-middle-class/ (accessed February 19, 2021).

Power

30 John Adams (1854). "The Works of John Adams, Second President of the United States: With a Life of the Author, Notes and Illustrations," p. 229
31 Mao, "Political power grows from the barrel of a gun." (Mao Zedong Quotes: https://www.azquotes.com/author/16154-Mao_Zedong).
32 https://duckduckgo.com/?q=tree+swinf+cartoon&t=newext&atb=v235-1&iax=images&ia=images&iai=https%3A%2F%2Fi.pinimg.com%2Foriginals%2F97%2F32%2F8e%2F97328e8b746969b31e0a56b8d55fe37a.jp

The Navy and Diversity

33 Task Force One Navy Final Report, https://media.defense.gov/2021/Jan/26/2002570959/-1-/-1/1/TASL%FORCE%20PONE%20FINAL%20REPORT.PDF

34 *Ibid.*
35 Christopher Rufo, "Ibram X. Kendi Is the False Prophet of a Dangerous and Lucrative Faith," *New York Post*, July 23, 2021.
36 Houston Keene, "Navy's Extremism Training Says It Is OK to Advocate for BLM at Work but not 'Politically Partisan' issues," Fox News, March 29, 2021.
37 DOD Military Demographics Report 2021, https://download.militaryonesource.mil/12038/MOS/Reports/2021-demographics-report.pdf.
38 National Center for Education Statistics, "Table 322.20 and Postsecondary Degree Fields Report," accessed September 23, 2023.
39 Task Force One Navy Final Report.
40 Email from senior DOD official indicates 2% of DOD workforce have concerns about racism.
41 Michael Lee, "Pentagon Report Finds about 100 Troops involved in Extremist Activity," Fox News, December 20, 2021.
42 Col (Ret) William F. Prince, "When a Diversity Conference Could Benefit from an Increase in Diversity," www.STARRS.us, September 19, 2023.
43 BG (Ret) Ernie Audino, "'Diversity' Is Not a National Security Imperative," www.STARRS.us, September 22, 2023.
44 CDR Philip Keuhlen, Task Force One Navy Final Report, "The Emperor's New Clothes, Redux," *Real Clear Defense*, December 6, 2021.
45 Col (Ret) Bing West, "The Military's Perilous Experiment," Hoover Institution, June 23, 2021, https://www.hoover.org/research/militarys-perilous-experiment

What Diversity, Equity, and Inclusion Is Doing to the Navy
46 https://www.americanthinker.com/articles/2023/12/what_diversity_equity_and_inclusion_is_doing_to_the_navy.html

How Social Justice is Killing the Military
47 2021 Demographics Report: Profile of the Military Community.
48 United States Census Bureau, August 21, 2021.

49 Department of Defense spokesman Jim Garamone, "Diversity, Equity, and Inclusion are necessities in the U.S. Military," *DOD News*, February 9, 2022.
50 www.STARRS.
51 *How to Be an Anti-Racist*, Ibram X. Kendi.
52 Report on the Fighting Culture of the United States Navy Surface Fleet by Lt Gen Robert Schmidle and RADM Mark Montgomery, July 12, 2021.
53 Market Watch, Mike Murphy, Gen Mark Milley, "I want to understand White rage, and I'm White," June 23, 2021.
54 *The Atlantic*, Ta-Nehesi Coates, "The Case for Reparations," June 2014.
55 https://www.americanthinker.com/articles/2022/03/how_social_justice_is_killing_the_military.html

How Implementing Diversity, Equity, and Inclusion Will Harm Readiness in the Armed Forces and Fail to Solve Anything

56 Christopher Rufo, Critical Race Theory: What it is and how to fight it, *Imprimus*, March 2021, https://imprimis.hillsdale.edu/critical-race-theory-fight/
57 Douglas Andres, "The Obama-Biden Military Purge," *The Patriot Post*, September 17, 2020, https:// patriotpost.us/articles/73508-the-obama-biden-military-purge-2020-09-17
58 *Ibid.*
59 Ta-Nehisi Coates, Guest Lecture to the Corps of Cadets, West Point, "The Case for Reparations," April 12, 2017, https://www.youtube.com/watch?v=oxQT3fQGJy4
60 Memorandum from the Executive Office of the President, September 4, 2020; An Executive Order on Combatting Race and Sex Stereotyping, September 22, 2020.
61 STARRS is Stand Together Against Racism and Radicalism in the Services, Inc., a 501(c)(3) non-profit whose mission is to educate our fellow Americans on the dangers of racist and radical ideologies infiltrating our military, in order to eliminate these divisive influences and maintain a unified and cohesive fighting force.

62 Memorandum for the Assistant Secretaries of the Navy, Chief of Naval Operations, and Commandant of the Marine Corps, May 3, 2021, Department of the Navy, "Diversity, Equity, and Inclusion Planning Actions"
63 Task Force One Navy Report, released January 26, 2021
64 Konstantin Toropin, Military.com, *Navy's Personnel Boss Says Getting Rid of Photos in Promotion Boards Hurt Diversity*, August 8, 2021, https://www.military.com/daily-news/2021/08/03/navys-personnel-boss-says-getting-rid-of-photos-promotion-boards-hurt-diversity.html
65 Coleman Hughes, *The City Journal*, "How to Be an Anti-Intellectual," October 27, 2019, httsp://www.city-journal.org/how-to-be-an-antiracist
66 CNO Professional Reading List, https://www.navy.mil/CNO-Professional-Reading-Program/
67 Barry Latzer, *National Review*, "Michelle Alexander Is Wrong About Mass Incarceration," April 4, 2019, https://www.nationalreview.com/magazine/2019/04/22/michelle-alexander-is-wrong-about-mass-incarceraton/
68 A report on the Fighting Culture of the United States Navy, July 21, 2021.
69 Lt Gen Rod Bishop, "Where's the Balance?" page 9.
70 Lynn Chandler Garcia, "I'm a Professor at the United States Air Force Academy. Here's Why I Teach Critical Race Theory," *The Washington Post*, July 6, 2021
71 Exploring Big Ideas: Diversity and Inclusion Reading Room Opens at Academy, February 19, 2021, https:// www.usafa.af.mil/News/News-Display/Article/2508630/exploring-big-ideas-diversity-and-inclusion-reading-room-opens-at-academy/
72 Liz George, "USAF Academy forces cadets to watch pro-Black Lives Matter video before attending school," *American Military News*, August 23, 2021, https:// americanmilitarynews.com/2021/08/video-usaf-academy-forces-cadets-to-watch-pro-black-lives-matter-video-before-attending-school/
73 Zachary Stieber and Jan Jekielek, *The Epoch Times*, Exclusive: "Space Force Officer, Punished After Denouncing Marxism, to Leave Military," August 17, 2021, https://www.theepochtimes.com/mkt_app/exclusive

74. Brittany Bernstein, *National Review*, June, 23, 2021, "Chairman of the Joint Chiefs Denies Military Has Gone 'Woke,' Says He 'Wants to Understand White Rage,'" https://www.nationalreview.com/news/chairman-of-the-joint-chiefs-denies-military-has-gone-woke-says-he-wants-to-understand-white-rage/

75. Ryan Morgan, *American Military News*, "West Point Cadets Forced to Attend Race Theory Seminars , including One Calling White Cops 'Murderers,'" April, 14, 2021, https://americanmilitarynews.com/2021/04/west-point-cadets-forced-to-attend-race-theory-seminars-incl-one-calling-white-cops-murderers/

76. *Judicial Watch*, August 10, 2021, "Judicial Watch Fights West Point Stonewall on Racist Propaganda," https://www.judicialwatch.org/investigative-bulletin/judicial-watch-fights-west-point-stonewall-on-racist-propaganda/

77. Tom Vanden Brook, *USA Today*, "Eight Cadets at West Point Expelled for Cheating, Over 50 Set Back a Year," April, 16, 2021, https://www.usatoday.com/story/news/politics/2021/04/16/west-points-worst-cheating-scandal-45-years-expels-eight/7121457002/

78. *Ibid.*

79. CITATION 1: Michelle Eberhart, *Pointer View*, "Author Coates Speaks at Zengerle Lecture," April 20, 2017, https://www.pointerview.com/2017/04/20/author-speaks-at-Zengerle-Lecture/ CITATION 2: Rich Lowry, *Politico Magazine*, "The Toxic World-View of Ta-Nehisi Coates," July 22, 2015, https://www.politico.com/magazine/story/2015/07/the-toxic-world-view-of-ta-nehisi-coates-120512/

80. Meghann Meyers, "The Defense Department Isn't Meeting Its Readiness Goals," *Military Times*, April 7, 2021, https:// www.militarytimes.com/news/your-military/2021/04/07/the-defense-department-isnt-meeting-its-readiness-goals-report-finds/

81. Lt Gen Robert E. Schmidle, USMC (Ret) and RADM Mark Montgomery, USN (Ret), "A Report on the Fighting Culture of the United States Navy Surface Fleet," April 2021, https://www.cotton.senate.gov/imo/media/doc/navy_report.pdf

82 *Ibid.*, page 7.
83 *Ibid.*, page 7.
84 *Ibid.*, page 8.
85 Notes of COL Pat Hueman, USA (Ret), taken 14 November 2021.
86 *Ibid.* Based on oral statement of former West Point cadets.
87 The Nuremberg Code specifies that people have the inherent legal right to protect themselves from medical experimentation. The Covid-19 vaccine ordered to be taken by military personnel is an experimental vaccine and has not been approved for use in the United States. Per CDC site December 2, 2021, "Pfizer-BioNTech Covid-19 Vaccine is **authorized for emergency use**," https:// www.fda.gov/emergency-preparedness-and-response/coronavirus-disease-2019-covid-19/comirnaty-and-pfizer-biontech-covid-19-vaccine.
88 Lt Gen Rod Bishop, "Where's the Balance?" pages 9-10
89 Statements of COL Pat Hueman, USA (Ret) and CAPT David Tuma, USN (Ret)
90 Thomas Sowell, *Discrimination and Disparities*, Basic Books, March 5, 2019
91 Virginia Kruta, "You Are Being Shaken Down": Author Shelby Steele Says Critical Race Theory is a Plot to Capture White Guilt," *The Daily Caller*, June 3, 2021, https:// dailycaller.com/2021/06/03/shelby-steele-critical-race-theory-fox-news/
92 Glenn C. Loury, *The Anatomy of Racial Inequality*, Harvard University Press, August 17, 2021
93 Wilfred Reilly, *Taboo: 10 Facts (You Can't Talk About),* Regnery Publishing, 2020
94 John McWhorter, "Woke Racism: How a New Religion Has Betrayed Black America," *Portfolio*, October 26, 2021
95 *Red, White, and Black: Rescuing American History from Revisionists and Race Hustlers,* Emancipation Books, 2021
96 *Black Eye for America: How Critical Race Theory Is Burning Down the House,* Be the People Books, 2021.
97 Julie Baumgardner, "The Millennial Success Sequence: First Things First," September 1, 2017, https:// firstthings.org/the-millennial-success-sequence/

98 Shannon Selin, *The Quotable Bonaparte—Nine of Napoleon's Most Memorable Quips Explained*, August 20, 2014, https://militaryhistorynow.com/2014/08/20/the-quotable-bonaparte-nine-of-napoleons-best-known-remarks/

Divide and Conquer: Radicalizing Military Education

99 Mounk, Y. (2018, October 5). "What an Audacious Hoax Reveals About Academia," *The Atlantic*. https://www.theatlantic.com/ideas/archive/2018/10/new-sokal-hoax/572212/

100 Beauchamp, Z. (2018, October 15). "The Controversy Around Hoax Studies in Critical Theory, Explained." *Vox*. https://www.vox.com/2018/10/15/17951492/grievance-studies-sokal-squared-hoax

101 Magness, P.W. (2018, November 28). "What the Hoax Papers Tell Us about the Decline of Academic Standards," The James G. Martin Center for Academic Renewal. https://www.jamesgmartin.center/2018/11/what-the-hoax-papers-tell-us-about-the-decline-of-academic-standards/

102 Hasnas, J. (2019, July 17). "The Diversity Distortion." The James G. Martin Center for Academic Renewal. https://www.jamesgmartin.center/2018/11/what-the-hoax-papers-tell-us-about-the-decline-of-academic-standards/

103 Goldstein, D., Grewal, M., Imose, R., Williams, M. (2022, November 18). *Unlocking the Potential of Chief Diversity Officers*. McKinsey & Company https://www.mckinsey.com/capabilities/people-and-organizational-performance/our-insights/unlocking-the-potential-of-chief-diversity-officers

104 Biden, J. (2021, June 25). *Executive Order on Diversity, Equity, Inclusion, and Accessibility in the Federal Workforce*. The White House. https://www.whitehouse.gov/briefing-room/presidential-actions/2021/06/25/executive-order-on-diversity-equity-inclusion-and-accessibility-in-the-federal-workforce/

105 U.S. Department of Defense. (2022). *Department of Defense Diversity, Equity, Inclusion, and Accessibility Strategic Plan, 2022-2023*. U.S. Department of Defense. https://media.defense.gov/2022/Sep/30/2003088685/-1/-1/0/DEPARTMENT-OF-DEFENSE-DIVERSITY-EQUITY-INCLUSION-AND-ACCESSIBILITY-STRATEGIC-PLAN.PDF

106 Cauthen, J. (2022, November 11). "The U.S. Naval Academy Is Adrift." The James G. Martin Center for Academic Renewal. https://www.jamesgmartin.center/2022/11/the-u-s-naval-academy-is-adrift/

107 The Heritage Foundation. (2023, March 30). "Report of the National Independent Panel on Military Service and Readiness." The James G. Martin Center for Academic Renewal. https://www.heritage.org/defense/report/report-the-national-independent-panel-military-service-and-readiness

108 United States Military Academy. (2017, October 9). *"Memo: New Curricular Addition Request for Diversity and Inclusion Studies Minor."* Department of the Army. https://www.dropbox.com/s/hc0kwzxc881rd8s/Diversity%20Minor%20Curriculum%20Committee%20%28Jan%202018%29.pdf?dl=0

109 United States Military Academy. (2020). *Class of '20 Inaugural Diversity & Inclusion Studies Minors.* Department of the Army. https://www.westpoint.edu/academics/academic-departments/behavioral-sciences-and-leadership/diversity-and-inclusion-studies

110 United States Military Academy. (2020). *Class of '20 Inaugural Diversity & Inclusion Studies Minors.* Department of the Army. https://www.westpoint.edu/academics/academic-departments/behavioral-sciences-and-leadership/diversity-and-inclusion-studies

111 United States Military Academy. (2023). *Mission.* Department of the Army. https://www.westpoint.edu/#:~:text=The%20U.S.%20Military%20Academy%20at%20West%20Point's%20mission%20is%20%22to,an%20officer%20in%20the%20United

112 United States Military Academy. (2023). *SS392 Politics, Race, Gender, Sexuality.* Department of the Army. https://courses.westpoint.edu/crse_details.cfm?crse_nbr=SS392&int_crse_eff_acad_yr=2017&int_crse_eff_term=2

113 United States Air Force Academy. (2023). *Gender, Sexuality, and Society.* Department of the Air Force. https://www.usafa.edu/app/uploads/COI.pdf

114 United States Naval Academy. (2023). *HE374 Gender and Sexuality Studies.* Department of the Navy. https://www.usna.edu/Academics/Majors-and-Courses/course-description/HE.php

115 Coatson, J. (2019, May 28). "The Intersectionality Wars." *Vox*. https://www.vox.com/the-highlight/2019/5/20/18542843/intersectionality-conservatism-law-race-gender-discrimination

116 Omokha, R. (2021, July 29) "I See My Work as Talking Back": How Critical Race Theory Mastermind Kimberlé Crenshaw Is Weathering the Culture Wars." *Vanity Fair*. https://www.vanityfair.com/news/2021/07/how-critical-race-theory-mastermind-kimberle-crenshaw-is-weathering-the-culture-wars

117 United States Naval Academy. (2021, March). *U.S. Naval Academy Diversity and Inclusion Strategic Plan*. Department of the Navy. https://www.usna.edu/Diversity/_files/documents/D_I_PLAN#:~:text=Our%20vision%20is%20that%20USNA,access%20to%20opportunities%20for%20success

118 United States Merchant Marine Academy. (2018). *Strategic Plan, 2018-2023*. Department of Transportation. https://www.usmma.edu/about/strategic-plan

119 United States Naval Academy. (2030). *Strategic Plan 2030*. Department of the Navy. https://www.usna.edu/StrategicPlan/goals.php#panel3GOAL3SustainandfurtherelevateUSNAastheNationsflagshipundergraduateeducationleadershipandcharacterdevelopmentinstitution

120 United States Naval War College. (2023). *U.S. Naval War College 2023 Women, Peace, and Security Symposium*. Department of the Navy. https://usnwc.edu/News-and-Events/Events/US-Naval-War-College-2023-Women-Peace-and-Security-Symposium#:~:text=The%20Naval%20War%20College%20(NWC,and%20Post%2DConflict%20Transitions.%22

121 Wilson, E. (2020, March). "The US Naval War College—The Navy's "Home of Thought." *Sea History*. https://nwcfoundation.org/wp-content/uploads/2020/10/SH170-NWC-by-EW.pdf

122 United States Congress. (1986). *Goldwater-Nichols Department of Defense Reorganization Act of 1986*. United States Congress. https://history.defense.gov/Portals/70/Documents/dod_reforms/Goldwater-NicholsDoDReordAct1986.pdf

123 United States Naval War College. (2023). *College of Naval Warfare Core Curriculum.* Department of the Navy. https://usnwc.edu/college-of-naval-warfare/Core-Curriculum

124 Spaulding, M. (2023, February 10). "DEI Spells Death for the Idea of a University," *The Wall Street Journal.* https://www.wsj.com/articles/dei-spells-death-for-the-idea-of-a-university-diversity-equity-inclusion-academia-college-hillsdale-new-college-of-florida-open-discourse-1d2ca552

125 West, B. (2021, June 23). *The Military's Perilous Experiment.* The Hoover Institution. https://www.hoover.org/research/militarys-perilous-experiment

126 Brady, P.H., Waltz, M. (2023, March 24). "The Military Should Reject DEI and CRT," *The Wall Street Journal.* https://www.wsj.com/articles/the-military-should-stay-colorblind-race-dod-army-navy-air-force-racial-preferences-dei-e64ead41

127 Glenn, H. (2020, July 6). "West Point Graduates' Letter Calls for Academy to Address Racism" National Public Radio. https://www.npr.org/sections/live-updates-protests-for-racial-justice/2020/07/06/887540591/west-point-graduates-letter-calls-for-academy-to-address-racism

128 Bindon, D., Askew, S., Schaeffer, J., Smith, T., Kehn, C., Lowe, J., Monaus, N., Salgado, A., Blom, M. (2020, June 25). "Policy Proposal: An Anti-Racist West Point." https://s3.amazonaws.com/static.militarytimes.com/assets/pdfs/1594132558.pdf

The United States: Most Racist Nation Ever

129 https://www.americanthinker.com/articles/2022/08/the_united_states_most_racist_nation_ever.html

DEI Means the Death of Military Professionalism

130 Reagan National Defense Survey. Ronald Reagan Institute. (2022, November). https://www.reaganfoundation.org/media/359970/2022-survey-summary.pdf

131 Report of the National Independent Panel on Military Service and Readiness. The Heritage Foundation. (2023, March 30). https://

www.heritage.org/defense/report/report-the-national-independent-panel-military-service-and-readiness

132 Cox, O., & Anders, C. (2022, December 1). "The U.S. Military Has a Politics Problem," *The Washington Post*. https://www.washingtonpost.com/politics/2022/12/01/us-military-has-politics-problem/

133 Reagan National Defense Survey. Ronald Reagan Institute. (2022, November). https://www.reaganfoundation.org/media/359970/2022-survey-summary.pdf

134 Reagan National Defense Survey. Ronald Reagan Institute. (2022, November). https://www.reaganfoundation.org/media/359970/2022-survey-summary.pdf

135 "Americans Divided on Whether 'Woke' Is a Compliment or Insult," *Ipsos*. (2023, March 8). https://www.ipsos.com/en-us/americans-divided-whether-woke-compliment-or-insult

136 Greene, J., & Paul, J. (2021, July 27). "Diversity University: DEI Bloat in the Academy." The Heritage Foundation. https://www.heritage.org/education/report/diversity-university-dei-bloat-the-academy

137 Report of the National Independent Panel on Military Service and Readiness. The Heritage Foundation. (2023, March 30). https://www.heritage.org/defense/report/report-the-national-independent-panel-military-service-and-readiness

138 Blackman, J. (2023, March 26). "Higher Education Faces an Inflection Point with DEI," *Reason*. https://reason.com/volokh/2023/03/26/higher-education-faces-an-inflection-point-with-dei/

139 Sailer, J. (2023, January 9). "How DEI Is Supplanting Truth as the Mission of American Universities," *The Free Press*. https://www.thefp.com/p/how-dei-is-supplanting-truth-as-the

140 Spalding, M. (2023, February 10). "DEI Spells Death for the Idea of a University," *The Wall Street Journal*. https://www.wsj.com/articles/dei-spells-death-for-the-idea-of-a-university-diversity-equity-inclusion-academia-college-hillsdale-new-college-of-florida-open-discourse-1d2ca552?mod=article_inline

141 Belkin, D. (2023, February 23). "Stanford Faculty Say Anonymous Student Bias Reports Threaten Free Speech," *The Wall Street Journal*.

https://www.wsj.com/articles/stanford-faculty-moves-to-stop-students-from-reporting-bias-anonymously-cbac78ed

142 Duncan, S. K. (2023, March 17). "My Struggle Session at Stanford Law School," *The Wall Street Journal*. https://www.wsj.com/articles/struggle-session-at-stanford-law-school-federalist-society-kyle-duncan-circuit-court-judge-steinbach-4f8da19e?mod=article_inline

143 Pinker, S., & Madras, B. (2023, April 12). "New Faculty-led Organization at Harvard Will Defend Academic Freedom," *The Boston Globe*. https://www.bostonglobe.com/2023/04/12/opinion/harvard-council-academic-freedom/

144 Pinker, S., & Madras, B. (2023, April 12). "New Faculty-led Organization at Harvard Will Defend Academic Freedom." *The Boston Globe*. https://www.bostonglobe.com/2023/04/12/opinion/harvard-council-academic-freedom/

145 Ulrich, M., Garcia, L. C., & Fitch, S. (2023, April 11). "Renewing Democracy Through Oath Education at the Air Force Academy." *War on the Rocks*. https://warontherocks.com/2023/04/renewing-democracy-through-oath-education-at-the-air-force-academy/

146 Ulrich, M., Garcia, L. C., & Fitch, S. (2023, April 11). "Renewing Democracy Through Oath Education at the Air Force Academy," *War on the Rocks*. https://warontherocks.com/2023/04/renewing-democracy-through-oath-education-at-the-air-force-academy/

147 Huntington, S. P. (1977). "The Soldier and the State in the 1970s," In *Civil-Military Relations* (pp. 27–27). essay, American Enterprise Institute.

148 Nix, D. E. (2012, March). "American Civil-Military Relations: Samuel P. Huntington and the Political Dimensions of Military Professionalism," U.S. Naval War College. https://digital-commons.usnwc.edu/

149 Lord, C. (2015, October). "On Military Professionalism and Civilian Control," *Joint Forces Quarterly*. https://ndupress.ndu.edu/Portals/68/Documents/jfq/jfq-78/jfq-78_70-74_Lord.pdf

150 Huntington, S. P. (2000). "Introduction: National Security and Civil-Military Relations," In *The Soldier and the State: The Theory and Politics of Civil-Military Relations* (pp. 2–2). The Belknap Press of Harvard University Press.

151 Department of Defense Diversity, Equity, Inclusion, and Accessibility Strategic Plan. Department of Defense. (2022, October). https://media.defense.gov/2022/Sep/30/2003088685/-1/-1/0/DEPARTMENT-OF-DEFENSE-DIVERSITY-EQUITY-INCLUSION-AND-ACCESSIBILITY-STRATEGIC-PLAN.PDF

152 Huntington, S. P. (2000). "Power, Professionalism, and Ideology," in *The Soldier and the State: The Theory and Politics of Civil-Military Relations* (pp. 83-83). The Belknap Press of Harvard University Press.

153 Bray, B. (2023, March 15). "The Navy Isn't Too Woke—It Is America." Center for International Maritime Security. https://cimsec.org/the-navy-isnt-too-woke-it-is-america/

154 Brooks, R. (2023, April 7). "How the Anti-Woke Campaign Against the U.S. Military Damages National Security." *War on the Rocks*. https://warontherocks.com/2023/04/how-the-anti-woke-campaign-against-the-u-s-military-damages-national-security/

155 Sailer, J. (2023b, January 27). "When Discipline-Specific Accreditors Go Woke." The James G. Martin Center for Academic Renewal. https://www.jamesgmartin.center/2023/01/when-discipline-specific-accreditors-go-woke/

156 Nathoo, Z. (2021, June 16). "Why Ineffective Diversity Training Won't Go Away," *BBC News*. https://www.bbc.com/worklife/article/20210614-why-ineffective-diversity-training-wont-go-away

157 The Editorial Board (Ed.). (2023, March 17). "The Tyranny of the DEI Bureaucracy," *The Wall Street Journal*. https://www.wsj.com/articles/judge-kyle-duncan-stanford-law-school-tirien-steinbach-dei-students-babc2d49

158 Ayas, R., Tilly, P., and Rawlings, D. (2023, February 7). "Cutting Costs at the Expense of Diversity." Revelio Labs. https://www.reveliolabs.com/news/social/cutting-costs-at-the-expense-of-diversity/

159 Foss, Nicolai J. and Klein, Peter G., "Why Do Companies Go Woke?" (November 23, 2022). Available at SSRN: https://ssrn.com/abstract=4285680 or http://dx.doi.org/10.2139/ssrn.4285680

160 Krauss, L. (2021, October 20). "How 'Diversity' Turned Tyrannical," *The Wall Street Journal*. https://www.wsj.com/articles/diversity-tyrannical-equity-inclusion-college-marginalized-race-11634739677

161 Garfinkle, H. (1977). "Introduction." In *Civil-Military Relations* (pp. 2–2). essay, American Enterprise Institute.
162 https://www.americanthinker.com/articles/2023/04/die_means_the_death_of_military_professionalism.htm

Task Force One Navy Final Report: "The Emperor's New Clothes," Redux

163 https://www.navy.mil/Press-Office/News-Stories/Article/2490996/task-force-one-navy-completes-report-to-enhance-navy-diversity/ (retrieved 8/22/20210
164 Task Force One Navy Final Report https://media.defense.gov/2021/Jan/26/2002570959/-1/-1/1/TASK%20FORCE%20ONE%20NAVY%20FINAL%20REPORT.PDF (retrieved 8/22/2021)(TF1N)
165 *Ibid.*, Appendix B, Task Force One Navy Charter. See "Purpose" and "Method."
166 *Ibid.*, p.6
167 *Ibid.*
168 "Ethnic Diversity Deflates Price Bubbles," December 30, 2014, *Proceedings of the National Academy of Sciences of the United States of America* https://www.pnas.org/content/111/52/18524 (retrieved 8/23/2021) (PNAS)
169 "Diversity Matters," McKinsey & Company, February 2, 2015 https://www.mckinsey.com/~/media/mckinsey/business%20functions/organization/our%20insights/why%20diversity%20matters/why%20diversity%20matters.pdf (retrieved 8/23/2021) (McKinsey)
170 PNAS p. 18524
171 *Ibid.*, p. 18525
172 *Ibid.*, pp. 18525-18526
173 *Ibid.*, p. 18524
174 *Ibid.*, p. 18525
175 https://www.mckinsey.com/about-us/overview (retrieved 8/23/2012)
176 EBIT: Earnings Before Interest and Taxes, a common measure of profitability
177 McKinsey, p. 1
178 TF1N, p. 6
179 McKinsey, p. 3
180 *Ibid.*, p. 2

181 *Ibid.*, p. 1
182 *Ibid.*, p. 4
183 *Ibid.*, p. 12
184 *Ibid.*
185 *Ibid.*
186 TF1N, p. 12 "Selectively" because if the goal were *truly* to mirror the demographic diversity of the nation as a whole, there would be critical TF1N discussion of the 17% "shortfall" in White enlisted members, 39% shortfall in females, and 6% "excess" in Black enlisted members (that dwarf the number of service members represented by the well-discussed putative "shortfalls" of Black and Hispanic officers), and there would be the identification of associated recommendations that are wholly absent. This is one of the clearest indicators of the moral and intellectual bankruptcy of the TF1N Final Report.
187 US Census, Annual Estimates of the Resident Population for Selected Age Groups by Sex for the United States: April 1, 2010, to July 1, 2019
188 US Census Bureau, Educational Attainment in the United States: 2015
189 US Census Bureau Releases New Educational Attainment Data, March 30, 2020
190 TF1N, pp. 12-17
191 See footnote 25
192 See footnote 25
193 TF1N, p. 103
194 See footnote 24
195 TF1N, p. 103
196 *Ibid.*, p. 104
197 *Ibid.*, p. 42
198 *Ibid.*, p. 108
199 Exec. Order No. 13583, 3 C.F.R. 13583 (2011), 76 FR 52845-52849 (2011)
200 Delgado, R and Stefanic, J, *Critical Race Theory, 3rd Ed.*, New York, NYU Press, 2017. "Seminal" because Delgado, along with Derrick Bell and Alan Freeman, is a cofounder of CRT.
201 *Ibid.*, pp. 167-186
202 *Ibid.*, pp. 8-11, 19-39

203 *Ibid.*, pp. 3-7. Delgado discusses the origins of CRT in critical legal studies and radical feminism, its growth with various identity variants, and its association with spin-off movements in education, philosophy, political science, and similar fields.
204 *Ibid.*, p. 3
205 *Ibid.*, pp. 25-26
206 *Ibid.*, pp. 8-9, 26-29
207 *Ibid.*, pp. 66-71
208 *Ibid.*, pp. 28-29
209 *Ibid.*, p. 27
210 *Ibid.*, pp. 103-104, 132-135
211 Ruf, Jessica, "Chief Naval Personnel Says Removing Promotion Photos Has Hurt, Rather Than Helped, Diversity Goals," *Diverse Military*, August 6, 2012 (retrieved at: https://www.diversemilitary.net/other-news/article/15076207/chief-naval-personnel-says-removing
212 5 U.S.C. 2301

Now Hear This
213 USNI News Fleet and Marine Tracker, August 28, 2023, USNI online
214 Navy Force Structure and Shipbuilding Plans: Background and Issues for Congress, April, 19, 2023, Congressional Research Service, https://s3.documentcloud.org/documents/23784604/navy-force-structure-and-shipbuilding-plans-background-and-issues-for-congress-april-19-2023.pdf
215 https://www.heritage.org/military-strength/assessment-us-military-power/us-navy, accessed September 3, 2023
216 ANALYSIS: "Shipyard Capacity, China's Naval Buildup Worries U.S. Military Leaders," January 26, 2023, *National Defense Magazine*, https://www.nationaldefensemagazine.org/articles/2023/1/26/analysis-shipyard-capacity-chinas-naval-buildup-worries-us-military-leaders
217 CNO Admiral Gilday stated in August of 2022, that "the biggest barrier to adding more ships to the Navy is the industrial base capacity." Congressman Grothman: "U.S. Naval Shipbuilding Capacity Must Compete With China & Russia," United States House Committee on Oversight and Accountability, Congressional Hearing, 11 May 2023,

https://oversight.house.gov/release/grothman-u-s-naval-shipbuilding-capacity-must-compete-with-china-russia/

218 China Naval Modernization: Implications for United States Navy Capabilities, May 15, 2023, Congressional Research Service

219 "China adds 22 ships to world's largest coast-guard fleet, Japanese news agency says," February 2, 2023, *Stars and Stripes*, https://oversight.house.gov/release/grothman-u-s-naval-shipbuilding-capacity-must-compete-with-china-russia/

220 Weapons System Sustainment: Navy Ship Usage Has Decreased as Challenges and Costs Have Increased, January 2023, GAO-23-106440

221 https://www.heritage.org/military-strength/assessment-us-military-power/us-navy, accessed September 3, 2023

222 "A Report on the Fighting Culture of the United States Navy Surface Fleet," Commissioned by Senator Tom Cotton, Congressman Jim Banks, Dan Crenshaw, and Mike Gallagher, December 2021

223 "The Navy finds major failures, starting with top officers, in a devastating ship fire, October 19, 2021, National Public Radio online, https://www.npr.org/2021/10/19/1047428110/the-navy-finds-major-failures-starting-with-top-officers-in-a-devastating-ship-f

224 "Long Chain of Failures Left Sailors Unprepared to Fight *USS Bonhomme Richard* Fire, Investigation Finds," October 19, 2021, *USNI News*, https://news.usni.org/2021/10/19/long-chain-of-failures-left-sailors-unprepared-to-fight-uss-bonhomme-richard-investigation-finds

225 Modly Resigns as Acting SecNav Amid Backlash Over Carrier Captain's Firing, April 7, 2020, www.Military.com.

226 Surface Warfare Tackles Persistent Problems as More Than Half of JOs Say They Don't Want Command, June 19, 2023, USNI online, https://news.usni.org/2023/06/19/surface-warfare-tackles-persistent-problems-as-more-than-half-of-jos-say-thy-dont-want-command

227 Navy Ford Class Aircraft Carrier Program, August 16, 2023, Congressional Research Service

228 "Nearly 100 Deaths, Half a Million Cases: The Toll from 3 Years of the Coronavirus Pandemic on the Military," March 28, 2023, Military.com, https://www.military.com/daily-news/2023/03/28/nearly-100-deaths-half-million-cases-toll-3-years-of-coronavirus-pandemic-military.html

229 "It 'keeps us awake': Navy leaders say sailor suicides are huge concern," January 11, 2023, Navytimes.com, https://www.navytimes.com/news/your-navy/2023/01/11/it-keeps-us-awake-navy-leaders-say-sailor-suicides-are-huge-concern/

You Decide
230 National Convention on October 27, 1964.

He Who Does Not Risk, Cannot Win
231 https://www.americanthinker.com/articles/2023/01/who_will_not_risk_cannot_win.html

Liberty Matters More
232 About Black Lives Matter, https://blacklivesmatter.com/about/
233 Rav Arora, "These Black lives didn't seem to matter in 2020," *New York Post*, February 6, 2021, https://nypost.com/2021/02/06/these-black-lives-didnt-seem-to-matter-in-2020/
234 Richard Brodie, *Virus of the Mind*, (Seattle, Integral Press, 1996).

A Sailor's Reflections on Race and the Navy
235 Brent Ramsey, *Armed Forces Press*, September 11, 2022. https://armedforces.press/shipmates-a-dying-breed
236 https://www.americanthinker.com/articles/2023/01/a_sailors_reflections_on_race_and_the_navy.html

The Navy's Misplaced Priorities Versus Core Navy Priorities
237 https://www.realcleardefense.com/articles/2020/06/11/the_middle_kingdom_rises_115372.html
238 https://nationalinterest.org/blog/reboot/revolt-admirals-how-navy-fought-truman-over-nuclear-carriers-179444
239 https://thefederalist.com/2023/05/08/like-bud-light-the-u-s-navy-steps-on-a-rake-with-drag-influencers/
240 Institute for Creation Research, Jonathan K. Corrado, Ph.D., P.E., "Matthew Fontaine Maury: The Father of Oceanography," October 27, 2022.

241 https://www.washingtonexaminer.com/policy/defense-national-security/naval-secretary-climate-change-top-priority

242 Annual Report on Suicide in the Military CY 2021 with CY21 DoDSER (1).pdf (dspo.mil); "Covid Deaths in U.S. Military Spike in Last Four Months Despite 96 Percent Being Vaccinated" (newsweek.com)

243 Heritage Foundation, Index of Military Strength Executive Summary October 22, 2022.

244 "The Navy has fired a dozen leaders but won't explain why" (nbcnews.com)

245 "Braithwaite to SASC: Navy Department in 'Troubled Waters' Due to Leadership Lapses, 'Tarnished' Culture"—*USNI News*

246 New details emerge about the 2020 *Bonhomme Richard* fire, ahead of censure of three-star (defensenews.com)

247 National Defense Authorization Act, Section 1092.

248 "Marines want 31 amphibious ships. The Pentagon disagrees. Now what?" (defensenews.com)

249 "Marine Corps No More?" | Hoover Institution Marine Corps No More?

250 "GAO Report on Columbia-class Submarine Program"—*USNI News*

251 "Report: Carrier *USS Ford's* Electromagnetic Systems Still Need Work" (maritime-executive.com)

Great Diversity Hoax

252 Burrow, M. (2023, May 8), "EXCLUSIVE: US Air Force Ran a Social Experiment to Graduate More Minority Pilots. It Didn't Go as Planned," *Daily Caller*, https://dailycaller.com/2023/05/08/air-force-social-experiment-diversity-minority-discrimination/

253 https://www.usna.edu/Diversity/index.php

254 https://www.archives.gov/milestone-documents/executive-order-9981

255 https://www.archives.gov/milestone-documents/executive-order-9981

256 Hanson, V.D. (2021) *The Dying Citizen*. Hatchet Book Group.

257 https://www.usna.edu/Diversity/staff/index.php

258 https://www.nytimes.com/2021/01/28/learning/whats-going-on-in-this-graph-diversity-in-professional-sports.html

259 https://www.usafa.edu/af-academy-offers-cadets-optional-diversity-and-inclusion-minor/
260 https://nces.ed.gov/programs/digest/d21/tables/dt21_313.10.asp
261 Hanson, *op. cit.*, p. 133

The Navy and Diversity, Equity, and Inclusion
262 American Indian/Alaska Native
263 Native Hawaiian/Pacific Islander
264 Improved Race, Ethnicity Measures Show U.S. Is More Multiracial (census.gov)
265 https://www.realcleardefense.com/articles/2023/07/18/the_navy_and_diversity_equity_inclusion_dei_966764.html#_ftn1
266 *Real Clear Defense*, Phillip Keuhlen, Task Force One Navy Final Report: "The Emperor's New Clothes, Redux," December 6, 2021.

Wokeness Is Antithetical to Military Service
267 Province, C. M., (1970, 2005), Northeast Kansas Korean War Memorial, http://www.pattonhq.com/koreamemorial.html

Danger Close: People's Republic of China
268 CIMSEC, "Ignorance Is Not Bliss," Brent Ramsey, March 4, 2020; *Real Clear Defense*; "The Middle Kingdom Rises," Brent Ramsey, June 11, 2020, CD Media, *A New China Strategy*, January 12, 2022.
269 Fox News, Yael Halon, Gordon Chang Warns China is Configuring Its Military to Kill Americans, September 7, 2020.
270 Michael Pillsbury, "The Hundred Year Marathon: China's Strategy to Replace America as the Global Superpower," March 16, 2016.
271 H. R. McMaster, Congressional Testimony, February 28, 2023.
272 US-China Economic and Security Review Commission Executive Summary, 2022.
273 Dr. Jamie Gaida, Jennier Wong Leung, Stephan Robin, Danielle Cave, Dannielle Pilgram, Australian Strategic Policy Institute, "Critical Tech Tracker-Sensors and Biotech Updates," March 1, 2023.
274 Senator Marco Rubio, "Made in China 2025," Report, February 12, 2019.

275 Brent Ramsey, "The Looming Taiwan Challenge for the US, Japan, and the West," CD Media, July 31, 2021 and "A New China Strategy," January 20, 2022.
276 Courtney Kube and Mosheh Gains, NBC News, "Air Force General Predicts War with China in 2025," January 27, 2023.
277 Heritage Foundation, Index of U.S. Military Strength 2023.
278 Dan Boyce, NPR, "Sexual Assault in the Military Increased 18%," March 12, 2023.
279 Stand Against Racism and Radicalism in the Services, www.STARRS.us.
280 Calvert Task Group, calvertgroup69.org
281 Restore Liberty—Enabling the Founders' vision of a free and fair republic (restore-liberty.org)
282 Center for Military Readiness, https://cmrlink.org/
283 Report by Senator Rubio and Congressman Roy, "How Political Ideology Is Weakening the American Military," November 22, 2022.

Connecting The Dots
284 Victor Davis Hanson, *The Dying Citizen*, (New York, NY: Hatchet Book Group, 2021), 16.
285 https://www.usna.edu/Diversity/index.php
286 Jack Beyer, "Military Service Academies Go Woke," *The Washington Free Beacon*, April 13, 2021, https://freebeacon.com › national-security › military-service-academies-go-woke
287 *Standage v Braithwaite, et al.* (Third Amended Complaint for Declaratory Relief, Civ. Action No.: 1:20-cv-02830-ELH (USDC, D.Md.) at 8, 24, 25)

In Their Own Words: "What Is Critical Race Theory?"
288 Keuhlen, P.J., Which United States Constitution? https://amgreatness.com/2021/10/17/which-united-states-constitution/, retrieved 7/26/23
289 Keuhlen, P. J., Task Force One Navy Final Report, https://www.realcleardefense.com/2021/12/06/task_force_one_navy_final_report_806508.html, retrieved 7/26/23

290 Executive Order 13583 https://obamawhitehouse.archives.gov/the-press-office/2011/08/18/executive-order-13583-establishing-coordinated-government-wide-initiativ, retrieved 7/29/23

291 Delgado, R & Stefanic, J, *Critical Race Theory, 3rd Ed*, New York, NYU Press, 2017. As my friend Dr. Scott Sturman, USAFA '72, reminds me, "All demagogues and radicals like Delgado, Marcuse, Hitler, Mao, and Marx inform the public years in advance of their intentions. When the ensuing turmoil occurs, the vast majority plead ignorance or claim the tyrant wasn't serious. Delgado was adamant about the control of language. Even the most mundane words and phrases are defined in the glossary section of his book *CRT*. That section fills a third of the book. His aim, along with Derrick Bell, is to silence critics and make sure in academia and the media and political arenas that their team always has home-field advantage.

292 Wood, Peter W, *1620: A Critical Response to the 1619 Project*, New York, Encounter Press, 2020. See also, Milke, Mark Ed, *The 1867 Project*, Canada, Aristotle Foundation, 2023, for a keen appreciation that the attack of the Crits on Western law and culture is international in scope.

293 See in particular the excellent short discussion of the Founding Fathers' strong views on the necessity of strong assimilation to the success of the American experiment at https://www.washingtonexaminer.com/michelle-malkin-assimilation-and-the-founding-fathers, retrieved 7/26/23

294 See discussion https://en.wikipedia.org/wiki/Economic_democracy#:~:text=Economic%20democracy%20(sometimes%20called%20a,includes%20workers%2C%20consumers%2C%20suppliers%2C, retrieved 7/29/23

295 Keuhlen, P. J., Task Force One Navy Final Report, https://www.realcleardefense.com/2021/12/06/task_force_one_navy_final_report_806508.html, retrieved 7/26/23

296 Keuhlen, P.J. USNA "CRT Smoking Gun," https://starrs.us/usna-crt-cp-smoking-gun-train-the-trainers/ retrieved 7/26/23. Note in particular that there is no pretense of academic balance on the USNA CTL Faculty Resource page. (https://www.usna.edu/CTL/Faculty_Resources/Diversity.php) i.e., no works representing other viewpoints on DEI by authors such as Dr. Thomas Sowell, Dr. John McWhorter, or Dr. Carol Swain.

297 https://www.c-span.org/video/?529425-1/superintendents-us-military-service-academies-testify-curriculum-admissions-standards, retrieved 7/26/23
298 https://www.foxnews.com/politics/diversity-biden-joint-chiefs-chairman-nominee-placed-dei-forefront-air-force-leader, retrieved 7/26/23
299 *Ibid.*
300 *Ibid.*

No Institution Is Safe: Thought Control in the Military

301 https://kaiserreich.fandom.com/wiki/Totalism
302 https://newdiscourses.com/2023/02/ideological-totalism-in-the-woke-cult/
303 https://wellcomecollection.org/articles/YrwNbxEAACcALMB0
304 https://science.howstuffworks.com/life/inside-the-mind/human-brain/brainwashing.htm
305 https://freedomofmind.com/robert-jay-liftons-eight-criteria-of-thought-reform-brainwashing-mind-control/
306 http://www7.bbk.ac.uk/hiddenpersuaders/blog/interview-robert-jay-lifton-on-brainwashing-and-totalism/
307 https://www.smithsonianmag.com/smart-news/teenage-brains-are-like-soft-impressionable-play-doh-78650963/
308 https://www.nea.org/resource-library/educator-rights-create-inclusive-classroom
309 https://www.nea.org/professional-excellence/just-and-equitable-schools/core-values
310 https://www.aft.org/resolution/dei-and-racial-justice-investments
311 https://csalateral.org/issue/2/manifestos-occupy-wall-street/
312 https://mcc.hu/en/event/the-long-march-to-the-institutions-1
313 https://manhattan.institute/book/americas-cultural-revolution-how-the-radical-left-conquered-everything
314 https://starrs.us/
315 https://thefederalist.com/2022/10/25/no-affirmative-action-in-the-military-doesnt-boost-national-security-it-erodes-it/
316 https://www.americanthinker.com/articles/2022/10/activist_generals_are_transforming_the_air_force_academy.html

317 https://emile-education.com/critical-pedagogy-what-is-it/
318 https://criticalrace.org/service-academies/united-states-naval-academy/
319 https://www.usna.edu/Diversity/
320 https://www.cpshr.us/resources/how-diversity-improves-organizational-performance
321 https://www.researchgate.net/publication/360939244_Why_Diversity_Programs_Fail
322 https://christopherrufo.com/p/diversity-programs-miss-the-point?r=l4ziw&utm_medium=ios&utm_campaign=post
323 https://www.youtube.com/watch?v=8IqPhXYuodo&t=5664s
324 https://www.theepochtimes.com/epochtv/is-dei-doomed-carol-swain-on-new-allegations-against-harvards-president-atlnow-5545204?utm_source=ref_share&src_src=ref_share&utm_campaign=mb-cc&src_cmp=mb-cc
325 https://www.researchgate.net/publication/326304028_Concept_analysis_of_impressionability_among_adolescents_and_young_adults
326 https://www.realcleardefense.com/articles/2023/09/25/the_united_states_air_force_academys_white_boy_2_981680.html
327 https://armedforces.press/has-dei-become-more-important-than-academics-at-the-air-force-academy/
328 https://www.realcleardefense.com/articles/2023/09/25/the_united_states_air_force_academys_white_boy_2_981680.html
329 https://www.cna.org/pop-rep/1998/html/4-commission.html
330 https://www.youtube.com/watch?v=_6xBiyidQ9g
331 https://www.heritage.org/defense/commentary/the-military-recruiting-crisis-getting-worse
332 https://armedforces.press/diversity-equity-and-inclusion-or-shipmateyou-cant-have-both/
333 https://www.bloomberg.com/opinion/articles/2023-07-04/us-military-recruiting-crisis-is-a-national-security-emergency
334 https://www.americanthinker.com/articles/2023/12/inside_the_woke_air_force.html
335 https://thefederalist.com/2022/10/25/no-affirmative-action-in-the-military-doesnt-boost-national-security-it-erodes-it/

336 https://www.wsj.com/articles/the-military-should-stay-colorblind-race-dod-army-navy-air-force-racial-preferences-dei-e64ead41

337 https://nationalinterest.org/blog/buzz/us-air-force-serious-decline-207728

338 https://www.newsweek.com/us-military-went-woke-time-make-some-changes-top-opinion-1849290

339 https://thelimegroup.com/wordpress/wp-content/uploads/2018/02/Critical-mass-and-tipping-point-in-change-efforts.pdf

340 https://legal-dictionary.thefreedictionary.com/Medical+Experimentation

341 https://www.youtube.com/watch?v=K1c26q0oxoE

The Constitution, The Officer's Oath, Core Values, and Admirals

342 www.navy.com, "Who we are"

343 *New York Post*, Chris Rufo, "What Critical Race Theory Is Really About," May 6, 2021.

344 *Real Clear Defense*, Task Force One Navy Report: "The Emperor's New Clothes Redux," May 6, 2021.

345 Frank Dobbin and Alexandra Kalev, *Harvard Publications*, "Why Diversity Training Does Not Work," May 21, 2021.

346 Col Michael Pefley, USAF (Ret) and CAPT Brent Ramsey, USN (Ret), "Racism in America Is Vanishing," CD Media, August 7, 2022.

A Path Forward for the Navy and the Nation

347 Sowell, T., (2021, July 23), "Top 20 Quotes from Thomas Sowell About Race," *The Burning Platform*, #10, https://www.theburningplatform.com/

348 *Ibid.*, #5

349 Carson MD, B, (2023, January 3), *Dr. Ben Carson Diagnoses American Schools' Greatest Ailment as "Loss of Vision,"* The Daily Signal, https://www.dailysignal.com/2023/01/03/dr-ben-carson-diagnoses-americas-greatest-ailment-as-loss-of-vision/

A Time of Challenge—Moving Toward the Light

350 Catoir, Fr. John, *Joyfully Living the Gospel Day by Day,* Catholic Book Publishing, 2001.

351 Sheen, Archbishop Fulton J.. *A Declaration of Dependence: Trusting God Amidst Totalitarianism, Paganism, and War,* Sophia Institute Press, 1941.

352 http://www.john-adams-heritage.com/quotes/

The Myth of an Apolitical Military: A Call to Action

353 Keuhlen, P. J., "Which United States Constitution?" *American Greatness,* October 17, 2021, https://amgreatness.com/2021/10/17/which-united-states-constitution/ retrieved 5/28/2023

354 https://boghossian.substack.com/p/woke-religion-a-taxonomy

355 Swain and Pierce, *The Armed Forces Officer,* NDSU, DOD, 2017 (https://bookstore.gpo.gov/products/armed-forces-officer)

356 Title VII Civil Rights Act of 1964 (https://www.eeoc.gov/statutes/title-vii-civil-rights-act-1964)

357 Sturman and Keuhlen, "No Institution Is Safe: DEI Thought Control in the Military, 2023" (https://brownstone.org/articles/no-institution-is-safe-dei-thought-control-in-the-military/)

358 Swain and Schorr, *Black Eye for America,* Be the People Books, 2021, pp. 69–78

Made in the USA
Columbia, SC
03 November 2024

45572771R00215